Fool
Proof

Fool
Proof

How FEAR *of* PLAYING *the*

SUCKER SHAPES OUR SELVES

and the SOCIAL ORDER—AND

WHAT WE CAN DO ABOUT IT

Tess Wilkinson-Ryan

HARPER WAVE
An Imprint of HarperCollins*Publishers*

HarperCollins books may be purchased for educational, business, or sales promotional use. For information, please email the Special Markets Department at SPsales@harpercollins.com.

FIRST EDITION

Designed by Bonni Leon-Berman

Library of Congress Cataloging-in-Publication Data

Names: Wilkinson-Ryan, Tess, author.
Title: Fool proof : how fear of playing the sucker shapes our selves and the social order-and what we can do about it / Tess Wilkinson-Ryan.
Description: First edition. | New York, NY : Harper Wave, an imprint of HarperCollinsPublishers, [2023] | Includes bibliographical references and index. | Summary: "Moral psychologist Tess Wilkinson-Ryan examines what she calls the "sucker construct," the fear of being taken advantage of-of being a sucker-and its outsized role in the shaping of our lives and our world"— Provided by publisher.
Identifiers: LCCN 2022038352 (print) | LCCN 2022038353 (ebook) | ISBN 9780063214262 (hardcover) | ISBN 9780063214286 (epub)
Subjects: LCSH: Social pressure. | Social influence.
Classification: LCC HM1176 .W55 2023 (print) | LCC HM1176 (ebook) | DDC 302/.13—dc23/eng/20220915
LC record available at https://lccn.loc.gov/2022038352
LC ebook record available at https://lccn.loc.gov/2022038353

22 23 24 25 26 LBC 5 4 3 2 1

To my parents,

Jane Ryan and Ren Wilkinson

Contents

Introduction

Often, waving good-bye after a visit or a family vacation, my mother will call after me, "Don't take any wooden nickels!" I think she's mostly joking, repeating something funny she heard from her own father, but it still has that parental urgency. *Please be careful. World, please be kind.*

In the American moral vernacular, we have a whole thesaurus for victims of exploitation. They are suckers (born every minute), fools (not suffered gladly), dupes, marks, chumps, pawns, and losers. Fool me once, shame on you; fool me twice, shame on me. Cultural stories about suckers abound too: the Trojan Horse, the Boy Who Cried Wolf, the Emperor's New Clothes, even Hansel and Gretel. If you believe that, I have a bridge to sell you. Don't take candy from a stranger. Don't go out with him; he only wants one thing. Don't be too credulous; the world will take advantage. The fear of playing the fool is not just a descriptive fact; it is a prescriptive theme: *Don't let that be you.*

What, exactly, is so bad about feeling suckered? Why is its sting so sharp, and why does it last so long? Most people will recognize the urge to recoil at the prospect of suckerdom, the potent combination of dread and disdain. But for a familiar, intuitive experience, one that we talk about socially, reinforce culturally, and ruminate on personally, the fear of being played for a fool gets remarkably little attention as a coherent phenomenon.

Actually, there are a lot of books and articles about how to *avoid* being scammed. This is not one of them. The questions at the center of this book are more curious about the psychological mechanisms and more skeptical about the cultural motivations for what gets called a scam and who gets called a sucker—accusations with surprisingly high stakes, for the self and for the social order. The goal here is not so much to spot the con but to renegotiate its meaning. Most of us are constantly navigating two sets of imperatives: how to be successful and how to be good. Fear of playing the fool not only gets in the way of both missions but often whispers that you can't do both. This book turns up the volume on that whisper so we can hear it clearly—so we can decide for ourselves when to listen and when to ignore, what to save and what to let go.

In the earliest days of the pandemic, my family would sometimes drive around on essentially pretextual errands, just to get out of the house. We were all tetchy, bewildered; the kids wore headphones in the back seat as I brought them to one or another unappealing destination. One afternoon I made them visit, I am not kidding, a really big empty parking lot. As we drove home along Broad Street in the early gray spring, my son, almost thirteen, said suddenly from behind me, "I think my biggest fear in the world is getting tricked into going on *Dr. Phil*." (I assume the sharp increase in YouTube time had exposed him to this phenomenon. I also believe and dearly hope that he has enough moral imagination that this is not literally his greatest fear, but I take his point.) His sister, thoughtful at eight years old, looked out the window at the traffic and said absently, "My biggest fear is hurting someone else." And there you have it. They neatly excavated two kinds of vigilance: Neither a sucker nor a scammer be.

The Mark

A mark is a chump—the pawn in a rigged game, the con artist's target. In mid-century lingo, a mark had to be "cooled out," because he would get so upset that he might throw the whole game into disarray. Being conned does feel terrible. You can probably make yourself cringe just thinking about playing the fool, a loser both literally and socially. The mark is a special kind of victim with a particular penchant for self-loathing.

It can be helpful to try a little thought experiment to see what makes the sucker experience singular. Imagine one afternoon you get an alert from your bank that there is a suspicious charge on your account for $20.50, put through one hour ago. The charge appears to be associated with a website called EZGamezzzz.com, a site you have never heard of, much less provided with your credit card information. The customer service agent from the bank tells you that it's certainly a fake company, likely deployed by hackers who test random digit strings and sometimes get lucky. The bank will refuse the transaction and you won't lose any money. Okay!

Now imagine slightly different facts. An hour ago you did give your credit card information out, and you did authorize a charge for $20.50, but not to EZGamezzzz.com. An affable young representative from an international children's fund was tabling outside your grocery store, asking people to give $20.50 as part of their End Child Poverty by 2050 campaign. It's not the sort of thing you would usually stop at, but the photos were compelling, and the guy at the table was pleasant, and naturally you are in favor of ending child poverty. When you get the alert from the bank, you realize something has gone wrong. A neighborhood LISTSERV confirms your suspicions, with a warning about the charity scammers outside of the supermarket. You thought you were helping impoverished children but you were paying a con artist. You give the bank an abbreviated version of this story and again the customer service agent agrees to block the charge. You won't lose any money.

In either case, the only material loss is a couple of minutes on the phone with a customer service agent, a little bit of hassle; the outcome to your bank account is the same whether you are a random victim or a cooperative dupe. But for most people these two scenarios feel very different. The hack is a minor annoyance, frustrating but not a cause for self-recrimination or shame. Getting conned is painful in a deeper and more complicated way, a lesson that sticks with its mark long after the bank records have been corrected. It implicates the sense of self (what was I *thinking*? How could I be so *stupid*?), and it also feels like a faux pas. Socially speaking, fools get little sympathy and a lot of scorn. There is something deeply disruptive about the feeling, or even the anticipation, of a fool's betrayal.

It makes sense that we want to avoid bad feelings. There is a reason you don't leave your door unlocked or your life savings on the front porch: things you value need to be protected. In the familiar cultural narrative, each mistake—wooden nickels, EZGamezzzz, Dr. Phil—is an opportunity for learning. Once burned, twice shy. But we talk less about the costs of being excessively wary. What lesson will you take from your brush with the scammer, and what shadows will it cast the next time you want to be generous or open-minded?

I am a professor of contract law with a graduate degree in psychology, so I am pretty comfortable with the idea that people care a lot about betrayals. Contract disputes are naturally about suckers, or at least about people who feel like suckers when their deals fall apart; nervous potential chumps occupy a lot of my day-to-day professional attention.

My original interest in fools, though, started from another side of my academic life. I finished law school in 2005, but I did not take the bar and become an attorney. While I was in law school, I had grown increasingly occupied with how regular people *perceive* their legal rights and obligations. Instead of getting a real job after graduation, I stayed in school for three more years and completed my PhD in experimental psychology. I was most curious about decisions that forced people to navigate legal rules that conflicted with their moral commitments. How would they solve those problems, and how would their solutions make them feel?

One of the things we had talked about a lot when I was a law student was the idea that people who break their contracts are not wrongdoers. Philosophical, legal, and economic discussions of contracts rarely couch breach in moral terms. By the time you finish law school, this impassive approach feels normal—but I don't know any non-lawyers who feel neutral about their contracts. Most people think breaching a contract is breaking a promise, and that's a big moral deal. This small observation about the difference between legal reasoning and ordinary moral reasoning became the core of my research and eventually my career, and it started with me just asking people: "How do you feel about breach of contract?"

I would send out different contracts hypotheticals to various subject pools, either online survey takers or law students or sometimes in-person samples from around the university. Usually I would ask them to read a short scenario about how the fictional character "Bob" had a contract to refinish someone's floors or something similarly mundane and had to decide whether to breach the contract when he got a much better offer from a new customer. The question I would pose to subjects was: If Bob breaks the deal, before any money has even changed hands, how much should a court require him to compensate his original customer for the breach?

For starters, people thought Bob should have to pay a lot. Sometimes people would enter an integer and then just fill in the rest of the response space with zeroes, so Bob had to pay, like, a googolplex dollars for failing to refinish some condo floors. Overall, subjects appeared to want not just compensation for the victim but also punishment for the breacher; they consistently chose money damages much higher than the law would normally allow.

I also got direct narrative feedback. At the end of each questionnaire, I had a standard free-response opening for "any additional comments or questions." I put it there so people had a place to tell me if the web link had been glitchy or if they found any of the questions confusing. Instead, a substantial fraction of respondents wrote in to express that they were enraged by just the *idea* of breach. They wrote lengthy free-form comments, along these lines: "This is what is wrong with America today. Your word used to be your BOND and now people do not respect each other" or "Bob is *BETRAYING* his customers." I encountered a lot of caps lock.

People felt like the breach of contract was a sucker's game, and they wanted the schemer to pay. I had offered a somewhat boring paragraph about a totally hypothetical floor-refinishing service, so I was impressed by the force of the outrage, and also pretty excited because I felt like I was onto something.

As is the norm for academics, I *was* onto something, but it wasn't really new. The psychology of the scorned sucker was a favorite topic of, among others, Erving Goffman, one of the twentieth century's most famous observers of social behavior. Goffman built a whole theory on a simple premise: people hate an interpersonal disruption—and the discovery of a scam is deeply, terribly disruptive.

Goffman was an unlikely hero of psychology, since he was actually a sociologist, and he had come to much of his theory by way of his deep interest in the theater. His "dramaturgical model of social life" drew from his childhood among amateur thespians in Manitoba, Canada. For his whole academic career, Goffman was preoccupied with the lengths people will go to to avoid embarrassment, to avoid ruining the scene by saying the wrong lines. He saw, though, how this impulse to smooth social dynamics makes people easy marks. They agree to go along, or nod politely instead of asking probing questions, and that's how they end up donating to a fake charity or buying a questionable time share. In 1952, still in graduate school, he published a short article on the sucker in his social world, a paper with the truly killer title "On Cooling the Mark Out: Some Aspects of Adaptation to Failure."

The essay was an allegory about the aftermath of the con, a story with three roles: an operator, a mark, and a cooler. The operator chooses his mark, the deal goes south—and then what? "In cases of criminal fraud," he began his article, "victims find they must suddenly adapt themselves to the loss of sources of security and status which they had taken for granted." Goffman saw a key tension: people hate to be embarrassed, it is embarrassing to be played for a fool—and also, the human condition is that people get played all the time. Humans must adapt, but how?

Enter the "cool-out." Goffman, clearly enjoying himself, wrote of the operator who "stays behind his team-mates in the capacity of what might be called a cooler and exercises upon the mark the art of consolation." The mark is offered an alternative narrative, an acceptable gloss on the facts that makes it possible to move forward. "For the mark, cooling represents a process of adjustment to an impossible situation . . ."

Goffman's cooler was an agent tasked with letting the marks down easy, cajoling them into taking the short end of the stick and keeping quiet about it. A cooler might offer a small consolation prize—a partial rebate or a coupon for next time—to take away some of the sting. Consoled, the marks could be persuaded to move on. Disruption over; system intact.

When Goffman talked about the cooling phenomenon, he had in mind something cynical and manipulative, which is understandable. But as a psychologist, I have a complementary take. The construct doesn't need to be pejorative, and it doesn't even need to be interpersonal. Sometimes I prefer to accept a bad deal and not have lots of feelings about it. Maybe I loan money to my cousin and she doesn't repay it; I could feel betrayed, but I'd rather feel proud that I could help. Sometimes I *want* to talk myself out of feeling suckered; sometimes cooling out is just my rejection of a frame that wasn't particularly helpful, or didn't vindicate my deeper goals.

Why should I feel like a mark if I don't have to?

The mark is not a natural category; it is contingent and malleable. Am I a chump if I let another driver merge in front of me at the very last second? Am I a sap if I give money to a person panhandling on the street? What about if I take a gamble on a dicey investment and it goes sideways? Maybe I'm the sucker, sure, but maybe I'm a laid-back road tripper, a compassionate donor, or an investor with a high risk tolerance—and maybe the uncertainty can offer a little breathing room.

These mundane dilemmas pop up all the time, and as a self-proclaimed expert on fools I attract a lot of stories about life's little scams. My sister Ivy, who has been listening to me talk about suckers for at least fifteen years, called one afternoon with a pleasingly literal example of self-cooling. She lives in Vermont, and she had taken a long bike ride with her husband and some friends. The ride turned out to be more than she had expected: she was in excellent shape for an oncologist, but the other riders were in excellent shape for triathletes. They coasted into a small town with a general store and stopped to buy snacks. Ivy was really thirsty and starting to feel light-headed.

"So we came to the store and it's not a regular convenience store: it turns out to be precious Vermont; like, they had house-made kombucha," she told me. "And I am just trying to get a regular Gatorade and they were charging, like, six dollars for it."

She was indignant. Sure, *tourists* might come to Vermont and be willing to pay $6 for Gatorade. *New Yorkers* might pay $6 for Gatorade and then get one of those white oval VT car stickers. She, on the other hand, knows better! (As her older sister, I am obligated to point out that she herself is not from there, either. I'm not sure what her claim is to some authentic old Vermont, aside from having grown up in Maine. She and I are squarely in the target demographic for Vermont cute. We love kombucha.)

"Anyway," she continued, "a Gatorade was literally worth one hundred dollars to me at that moment. I thought of you and was like, 'What am I doing???'" Thus cooled, she bought the drink and rode home.

The Stakes

Sucker dilemmas are more than just Gatorade and $20 scams. What Goffman understood was that being played for a fool means "the loss of sources of security and status." Americans often struggle to reckon openly with questions of social status and hierarchy; it feels embarrassing and shallow, and counter to an egalitarian narrative we have about ourselves. But suckers, chumps, stooges, and marks are canonically low-status—and status threats feel existential. The accusation that you have played the fool is serious: you have debased yourself, and, worse, cooperated in your own social demotion. Whether the scam is particularized and clear-cut or diffuse and structural, it feels terrible to look around and wonder, *Wait, am I the fool here?*

That dawning fear can be an unintentionally powerful motivator across larger-scale political and social dynamics. As rhetoric, it might be overt—an accusation or an insult—but often a sucker threat just lies in wait, more subterfuge than open warfare. Because the fear it inspires is both potent and covert, it is ripe for weaponization. Indeed, the deployment of sucker threats should feel very familiar to observers of American politics in the first part of the twenty-first century, when we all got a scared-straight-style education in sucker-phobic populism. The furious refusal that neither he, nor America, would play the fool was one of Donald Trump's most consistent political values.

One of his favorite provocations on the campaign trail was to recite the lyrics to "The Snake," a song written by the singer and civil rights activist Oscar Brown Jr. Although the song was originally written as a repudiation of racism, Trump appropriated the fable within it for a very different message. In the song, a woman finds a cold, injured snake, who pleads with her to take him in. She agrees, and he promptly bites her. As she is dying, the snake chides her: "You knew damn well I was a snake before you took me in."

Extending a hand to children, or asylum seekers, or the huddled masses makes you a sucker, not a saint. Trump recited the song as a cautionary tale: Watch out for the snakes among you. But it's worth puzzling for a minute over the particular snakes he was talking about. They were not criminal fraudsters or exploitative bosses; they were poor, desperate families, usually with children, asking for humanitarian asylum. They had minimal economic or political power and no material means. They would seem to be an unlikely threat. In fact, the focus on scams from the weak is key to the rhetorical move. The hustles that people are most afraid of, and most likely to call out, are those that threaten the status of the mark and imperil the societal status quo.

The themes that play out in the chapters of this book transcend scams and dupes; they are about social power and moral agency. The sucker rhetoric has a function: to threaten the status hierarchy. It also has a power: to make people react to protect that hierarchy by quashing the threat in one way or another. Because this sucker trigger is easy to pull and, once pulled, hard to ignore, it has profound consequences for cognitive processing and social cooperation. By invoking the duplicitous snake, Trump could make the compassionate instinct feel silly or naïve.

Fear of playing the fool dictates whom we are willing to trust and whom we hold at arm's length. It keeps our eyes on the hierarchy and regulates allocation of power and status. As a phobia, it is structurally similar to other fears; my fight-or-flight response when I see a bear might involve more running, but it's that same set of instincts that explains how people respond to the looming grift. The "fight" is intuitive, and it shows up everywhere from economics games played in labs to intimate partner violence to armed conflict. Threatened, the would-be sucker retaliates to thwart the attack on the social order.

Almost more invidious, though, is when the fearful mark reacts with flight, or avoidance, instead. It's not as dramatic looking as fighting back, but when I am too scared to trust or too skeptical to take a leap, those choices also have serious ramifications. Refusal to engage can mean turning down opportunities, withdrawing cooperation, or resisting the prosocial urges of generosity and compassion. That apprehensive tendency to retreat has social and political consequences, too, from healthcare to welfare to immigration.

The recurrent psychological themes here—operators, marks, status, power, agency, avoidance—implicate some familiar political debates and some buried cultural narratives. A scan of our society shows fools' fables everywhere. As an analytic lens, the sucker dynamic shapes the social construction of gender; the implications of dominance and weakness, credulity and savviness, are not parallel for men and women. It pervades racial animus and insists on racial hierarchy; it tells us whom to believe, whom to fear, and whom to scorn. Themes of hucksters and dupes reverberate through racist tropes, which often leave little daylight between the gleeful claims of intellectual inferiority (suckers) and the dark warnings that "they" are coming for white dominance and white money (connivers). The salience of the scam affects how we perceive social progress: Are we leveling the playing field or creating "welfare queens"? In turn, a heightened fear of sucker threats from marginalized groups has the perverse consequence of making those who hold the least formal power the most relentlessly subject to suspicion and surveillance, as audited earned-income tax credit filers, people ticketed for "driving while Black," and videotaped warehouse workers can tell you.

Conversely—and perversely—some scams are fundamentally self-cooling, at least in our economic system. There are a lot of cons that we have to let slide, because living in this world requires taking a lot of raw deals. It should not be surprising that when a hustle reinforces the status quo, we tend to call it something else. When you get hazed to join a fraternity, it's tradition. When Jeff Bezos takes home billions of dollars in a global pandemic, he's a genius. When banks charge elaborate overdraft fees, that's just business as usual. Perversely, the fear of playing the fool can make it really uncomfortable to call exploitative systems what they are: If the meritocracy is a scam, what does that make me?

The Goal

Should I share? Should I trust? Should I take this risk? If you ask people about their values in complex situations—economic, moral, and social dilemmas—their answer will be often be something like *integrity*. What choice vindicates my integrity? But that's a hard problem: integrity is complex and resource-intensive. Should I donate to end child poverty? Should I lend money to my cousin? If I have explicit moral values, and I have familial obligations, and the facts are complex, and the consequences are probabilistic—the answer is probably unclear. Most moral quandaries leave a lot of room for error.

By contrast, it can be disarmingly easy to figure out who's on top, who's putting what over whom. Like most people, I see the status dynamics in any room at a glance. (After four decades of social feedback, I can guess who will remember my name or save me a seat, and I can figure out who wants attention and who is going to get it without trying.) Psychologically, without even knowing it, I may replace the harder question with the easier, determining what will increase my status rather than what will redeem my moral commitments. Asked for a loan, I demur; I don't want people thinking I'm a dupe. Asked for forgiveness, I am churlish; I can't let you take advantage of my good nature.

Or maybe not. Whether your value system is centered on the Golden Rule, or vulnerability, or duty, or something else, there's a certain moral empowerment in facing the sucker head-on. We can cut the fearsome threat down to size, taking it seriously yet treating it as one variable among many, not a secret trigger of doom and degradation. We can perceive, and then defuse, its function as a weapon against social progress. The promise of reckoning with the sucker's fear is the possibility of expansive moral agency.

There is a real, well-documented power in accounting explicitly for our implicit fears; it is the basis of techniques like cognitive-behavioral therapy, mindfulness exercises, or even cost-benefit analysis. The threat of being the sucker tends to be slippery and

amorphous, suffusing our thinking but not quite announcing itself. This makes it really hard to reason with integrity, or even
consistency, through complicated dilemmas. Sometimes, though,
a pattern emerges, and suddenly a cacophony of social facts rear-
ranges itself into something with a social meaning. It's not just cha-
otic dread or inchoate shame: there is a scam in the offing. What
we do when the sucker's game comes into focus is up to us.

This book starts by just laying out the definition of a sucker's
game and the psychology of why it is such a potent construct. I
trace the fear of playing the fool through studies from across
the academy—psychology, sociology, economics, and even
philosophy—to show a set of predictable behavioral patterns that
not only explain individual human reactions but also society-level
conflicts and prejudices. The book ends with a consideration of
what it takes to cool out, and what we stand to gain.

In the winter of her fifth-grade year, two years into pandemic life,
my daughter played in a local basketball league for nine- and ten-
year-old girls. There was something about the experience of that
season that everyone—the kids, the families, the coaches—found
practically euphoric. It was hardly optimal basketball conditions;
the girls played in masks, pulling them down only to take sips of
water between periods. But they didn't really care, and for the
parents—it had just been so long since we had seen our kids out
in the world in this way, and it felt like a revelation. We sat in the
stands in KN95s, glasses fogged, giddy with the vicarious physical
pleasure of it. We brought our teenage son to the playoffs, and he
got it, too. When his little sister took a foul shot and actually made
it, he jumped up and pumped his fists and then laughed so hard, he
had tears in his eyes.

Half of the girls were new to basketball. As we watched game
by game, they were figuring out the rules and the norms in real
time. The referees were blessedly lax—any serious attention to

traveling, or double dribble, would have ground the entire operation to a halt—but every once in a while someone would get called
for a foul. In the early games, the offending player would often
apologize, to her coach or the ref, for breaking the rules. It was
a charming category error, hard to explain, although the coaches
tried. Fouling isn't really cheating, and it's breaking a rule that you
are sort of expected to break, but you also should try not to?

At the same time, our team's coach was trying to help his players
see the other side of this coin, too. Kids get really frustrated when
a game seems unfair. The bench would buzz with complaints that
the refs weren't calling pushing or double-teaming (not allowed at
this age) or whatever. One of the coach's lessons, which I get is a
standard youth sports thing, was just: *That's for me to worry about,
not you; you can let that stuff go, and you should*. My daughter loved
this. She took the advice seriously and let it free her. Are they being nice? Is this fair? Did they follow the rules? You don't have to
worry about that! After she defended the league's tallest player in
a very physical matchup, she commented on the way home, "You
know one thing that my opponents don't expect about me is I don't
hold a grudge. Also, did you know she follows me on TikTok
now?" Her team had just lost the last playoff game, 15 to 16.

The point is, you *can* feel cheated, but you don't have to. The
question is not whether threats exist but which ones deserve your
attention. The prospect of playing the fool doesn't have to feel existential. Maybe, I hope, the ability to recognize the sucker dynamic
is the ability to decide what risks are worth taking, what relationships are meaningful, when to share, and when to protest—in
other words, how to live with integrity in a sucker's world.

Fool
Proof

CHAPTER 1

The Fear

M y maternal grandfather, Luke, used to say that he was born in "aught eight." Nearly a century later, he was living with my grandmother in their small house in Concord, New Hampshire, when their children began to badger them. Luke and Yvonne were well into their ninth decades and their kids—six daughters and a son—were increasingly concerned about their health and safety, taking notice of the new mobility limits and creeping cognitive decline. Luke's body was giving him more trouble than his mind; my grandmother was spry enough but recently diagnosed with the first signs of Alzheimer's dementia. They still lived independently, and my grandmother did the driving.

"She's the body and I'm the brain," he would say gruffly.

My grandmother was struggling to remember things. Her kids discovered a partially opened tin can of tuna fish, warm and fetid, in the microwave. Laundry had grown mysterious. She had once moved her ironing board to the living room, working through baskets of shirts in front of the televised McCarthy hearings. Now she might answer the door wearing two or three men's sweaters hanging over her slacks.

On a visit my parents found her stymied by the mechanics of her kitchen phone, a literal lifeline. My father sat with her and picked it up. "Hello? Yvonne, it's for you," he said cheerfully into the dial tone, handing her the receiver. The role play tweaked some muscle memory, and she held it to her ear, relieved.

My mom brought her on outings, once to the L.L.Bean store, where they sat outside the changing rooms and regarded one another.

"Mom? Are you hungry?" asked the daughter.

"Not hungry. Just suspicious," answered the mother.

Luke and Yvonne did not have a next step in mind, although Luke threatened that maybe they would just "go live with Helen." His sister Helen had recently celebrated her 102nd birthday.

My grandfather had failed his driver's test when his fused vertebrae made it impossible to turn his neck to check for cross traffic. Instead, he would sit in the passenger seat and yell commands to my grandmother, who would mostly comply in a timely fashion. With the road safety writing on the wall, my mother and aunt were appointed from the sibling consortium to escort Yvonne to the doctor, where she was told—gently, I hope—that their two-body piloting system was no longer safe. Arrangements for drivers and deliveries were made, and my mother promised she'd be back to visit very soon.

This is the background against which my mother pulled into Concord a week later and learned that she was suspected, by her parents, of conspiracy to commit elder fraud. She arrived at the house and found the door unusually locked, my grandparents glowering from inside. She asked them to let her in. Luke shouted that they were onto her—the entire town was onto her. "It's in the papers, Jane," he yelled through the door.

It took some work to discern what plot, exactly, she was suspected of, but in the end she pieced it together. My mother and her sister were accused of seducing a local physician—my grandmother's doctor. Thus plied, the doctor had been willing, on this remarkable account, to join in the falsification of his patient's health records and the fraudulent deprivation of her driving privileges.

Although I think any doctor should be so lucky as to land a romantic escape with my mother and aunt, the sisters were somewhat unlikely seductresses, both psychotherapists with grown children and long-term marriages. There were a lot of plot holes in the story, too—what "papers" did they get in Concord that would cover this sort of affair?—but what united and grounded my grandparents was the swamp of suspicion that mired all of their inferences. If their daughters consulted with the doctor, it was a conspiracy; if they couldn't drive anymore, it was a nefarious plot. Theirs was a confabulation that reflected more vibe than fact.

During this period they also grew wary of the delivery drivers who would bring them Meals on Wheels. They suspected a home health aide of stealing things from the kitchen. They were rigid with vigilance, spotting a con artist in every visiting nurse and a scam in each act of filial care. Their paranoia was pathological, a classic presentation for early dementia. (Although my grandfather did not himself have Alzheimer's, he was generally inclined to extreme skepticism and thus well-suited to this folie à deux.)

Most of us will never have clinical paranoia, and yet the dread—"not hungry, just suspicious"—is deeply familiar. We have all been gripped with that sudden fear, the warning prickle of a scam on the horizon.

The fear of playing the fool is a cultural trope and a human experience that almost everyone understands. At one extreme the fear is paranoia—irrational and even diagnostic of mental illness, like my grandparents'. But we are also well acquainted with the other version of sucker's wariness, the honorable kind; even my grandparents' self-protective instincts were clearly rooted in straightforward values of prudence and thrift. They had lived their lives on cautions about wooden nickels, fools and their money, and caveat emptor. Until it was called paranoia, their wary stance was entirely sensible, even prescriptive. You *should* be afraid of people taking advantage of you.

The two extremes of suckerdom, paranoia and prudence, present a clean dichotomy, but that's a dramatically incomplete picture, because the real fools' dilemmas happen in the squishy middle. Am I a sucker if I trust a stranger? What if I invest in my friend's dubious startup or report every tip on my tax filing? Most of the time, fear of being duped is neither clearly irrational nor clearly rational. Everyday life is full of opportunities and fool's games—and no warning labels.

The systematic study of this lurking sucker's fear was only recently articulated as a coherent, distinct psychological phenomenon. The experimental psychologists Roy Baumeister, Kathleen Vohs, and Jason Chin coined the term "sugrophobia," joining the Latin roots for "sucking" and "fear" tongue-in-cheek (get it?), in 2007. In an article called "Feeling Duped: Emotional, Motivational, and Cognitive Aspects of Being Exploited by Others," they proposed an overarching theory of the sucker: that the fear of being scammed is a unique human experience, and we can plot its psychological triggers and consequences. What raises those hackles? A cooperative venture, the risk of betrayal, the premonition that you might sign up and be left holding the bag.

Sitting Ducks

To see how context clues set off the sugrophobic antennae, consider an early experiment in social psychology. Sam Gaertner, then a graduate student at City University New York, used two fraught sites for the wary—the telephone and the highway—to raise his subjects' sucker hackles.

It was 1971, and under Gaertner's protocol a random sample of Brooklyn voters received a phone call from a stranger. The stranger, audibly relieved that the call had been answered, shouted, "Hello. . . . Ralph's Garage. This is George Williams. . . . Listen, I'm stuck out here on the parkway, and I'm wondering if you'd be able to come out here and look at my car?" None of the Brooklyn voters who got this call were running an auto shop, so they would

each say something like "Uh, sorry, wrong number." But the caller
would persist according to a script:

> Listen I'm terribly sorry to have disturbed you but listen—I'm
> stuck out here on the highway, and that was the last dime I had!
> I have bills in my pocket, but no more change to make another
> phone call. Now I'm *really* stuck out here. What am I going to
> do now? . . . Listen . . . do you think you could do me the favor
> of calling the garage and letting them know where I am? I'll
> give you the number. They know me over there.

A dime for a call from a pay phone, the numbers drawn from a
public phone book, no cell phones at all—it's hard to imagine. (I
bought a used Subaru when I was a senior in high school, when
we had seen car phones on TV but not really in person yet, at least
not in Maine. Even I of highly cautious maternal oversight was
allowed to drive country roads at night with an old car, new skills,
and zero backup plan. Now the idea is so harrowing that my own
memory seems implausible.)

But in 1971, George Williams's story was plausible and even familiar, though strictly speaking false, since that was not his name
and the number he gave did not connect to Ralph's Garage. Instead it went to a receptionist who would take the caller's relayed
information and enter it as data, in this case for Sam Gaertner's
study of the psychology of helping behavior.

Thirty years later, in suburban North Carolina, a different team
of psychologists was offering unsuspecting subjects another suspicious story, although this time it was enticement rather than en-
treaty. Research assistants set up a table in a shopping mall food
court with a large sign reading: "Free Money! $1 Bills Available
Here!" and just counted: How many approached and how many
walked by? (In their original sugrophobia paper, Baumeister,
Vohs, and Chin cited this very study as a classic sucker's dilemma.)

These two studies were ostensibly about very different aspects
of human psychology. The original Ralph's Garage research

actually had a pretty complicated theory of race and political ideology, trying to understand whether helping behavior would change depending on the race of the caller and the age and political party of the target.* The "Free Money!" study was part of a suite of experiments about the psychological "power of 'free.'" In both cases the authors were leveraging their intuition that subjects would be tempted but wary, that their choices would be manipulable in that squishy middle. Even if you don't know what these studies are about, you can put yourself in the position of the subjects and imagine their decision process. *Should I do this?* leads naturally to *What if this is a scam?*

For myself, I can feel my way through the decisions in the hypothetical. How would I have decided what to do if I had gotten that call from George Williams? I like to think I'd start with a little cost-benefit analysis. What are my goals, and what does it cost to meet them? In this case, the moral calculus looks pretty easy. I find it gratifying to be helpful. Interpersonal generosity is part of my explicit value system. Meanwhile, my time is valuable, but it is not *that* valuable. Certainly I have wasted many minutes today much as I do every other day. So I prefer to believe that I would have taken a moment and called Ralph's Garage if I had gotten that request. Even as I write that, though, I can feel the frisson of doubt that might have stopped me—and I would be in good company. Those random Brooklyn voters hung up or refused to call the garage between one-quarter and one-third of the time, although all that was at stake was a twenty-second local call.

And the free money? I might reason that, as a matter of my baseline preferences, I usually prefer more money over less money. I would pick up a dollar on the street, and I would accept a dollar's

---

* It did. Gaertner had different research assistants play the role of George Williams, and he knew from pilot testing that targets could reliably identify the race of the callers by voice alone. Black callers got more refusals and more hang-ups; older and more conservative targets favored white callers over Black callers. The implicit racism is no coincidence; racial bias manifested in sugrophobic choices is a pattern I consider in more depth in chapter 5.

worth of free samples at the grocery store, so it seems like I would take free money if that was the sample on offer. But like almost everyone else at the mall that day, I would probably walk right by a "Free Money!" table. And in fact, more than *90 percent of people* avoided their table. Later they tried offering up to $50, and even then fewer than one in four people stopped to inquire.

Anyone who has ever been subscribed for years to a service after they accepted the "free trial" is justifiably suspicious that most offers of free money are not free at all. As kids, my sister and I filled out a newspaper offer form to get twelve CDs for a penny from the Columbia House Music Club. All you had to do was agree to let them send you one album of their choice every month for the rest of the year—refundable, of course, if you filled out the return form, and packed it up, and mailed it back within two weeks of receipt. This is how we ended up knowing all the lyrics to every song on the Right Said Fred debut album. If I saw the "Free Money!" table, I wouldn't want to be the sucker who politely sits through a sales pitch or feels pressured into giving them my email or signing up for their newsletter. As it turned out, though, there was no scam. They were just handing out dollars to anyone who approached the table and asked.

On the one hand, Ralph's Garage and "Free Money!" are both deeply familiar dilemmas. Who among us hasn't ignored a dubious plea for help or rejected an offer because it was too good to be true? On the other hand, it's not clear that the fear of being duped was a helpful emotion for those particular subjects. The costs of engagement were intentionally minimal: you walk away without your free money if the table is a scam; you call someone you didn't need to call if Ralph's Garage doesn't exist.

Of course, the real cost was never the money, the time, or the hassle. The price of engaging in a scam, even briefly, is the psychological cost of confronting the self as fool. Many of us *want* to call the garage just in case it might help, and we would want to check out the free money and see if it's legit. We demur because we are scared. We are afraid that if we take a risk and get it wrong, we are going to feel foolish, and that feeling is so bad that it's worth avoiding.

When Vohs, Baumeister, and Chin wrote about sugrophobia fifteen years ago, they identified an important state of the science: they pointed out that psychology knows a lot about the feeling of being duped. There are deep literatures on its constitutive elements like betrayal, regret, shame, and social emotions, among others. And yet, we almost never talk about the fear of being duped as a coherent construct—as something that drives our decisions and behaviors. It's true that the cognitive and emotional life of a potential sucker (that's everyone!) is complicated, but it can and should be dissected to answer the key questions: Why do I hold back when I want to be generous, hang up on the distressed caller, forgo the attractive opportunity that might be too good to be true? What *exactly* are we so afraid of?

The very concept of sugrophobia as a literal phobia suggests that the fear of playing the sucker is less like rational caution and more like the kinds of things on your personal deep-fear list. For me: snakes, ticks, and people being mad at me. These are encounters, even imagined encounters, that I experience as Really Bad whether or not they have any material consequences for my health, wallet, or social life. I am truly phobic about snakes. In theory, the crux of that fear is a fear of snakebites, which would suggest, for example, that I would be all right with defanged snakes or snakes in cages. I assure you I am not; the mere thought of being in a room with a caged snake elicits a deep, visceral dread. The fear of playing the fool is similar. The prospect of being a sucker makes people fearful enough to avoid small risks and minor affronts, even when their own conscious calculations would tell them that there's nothing to be afraid of. That's probably no big deal when all I'm missing out on is the Reptile House at the zoo, but it matters more when it warns people off from relationships or opportunities or experiences they really value.

Being suckered taps into two unusually aversive human conditions: regret and alienation. By coincidence or cultural conditioning or evolutionary adaptation, the pain of self-recrimination (regret) and the pain of social isolation (alienation) are each capable of sparking the kind of outsized misery that people will take pains to avoid.

The Cringe Reel

I have a mental tape I can play, especially around three o'clock in the morning, that's like a life highlight reel, but instead of highlights it's just regrets. There is a lot of material to sift through, but my brain has a viciously precise search engine that permits it to locate any regret by date or theme instantaneously. Being fooled is of course not the only regret trigger—unfortunately it's one among many—but it is an incredibly reliable regret trigger. One of the things that is so excruciating about being a sucker is that you have this mental footage of yourself, blithely agreeing to your own downfall, that threatens to play on repeat. Sucker's games are regret factories. You can't be a sucker unless you cooperate, and the road to self-recrimination is paved with naïve cooperation.

At the start of this book, I compared the reaction to being randomly targeted in a data hack with the reaction to giving money to a fake charity. Both cases are annoying for a few moments and then end up okay, but one of those scenarios comes with a cringe reel and the other does not. If I get a random bank alert because of a hack, for example, I am not regretful. I'm annoyed and wish the hacker had chosen someone else's number, or a different profession; overall, though, I am focused on the behavior of the hacker. There is nothing for me to feel badly about. Regret applies to the choices I make, and here the only choices are the criminal choices of the hackers themselves.

In the charity scam, however, I am focused on my poor judgment. The charity scammers are no less to blame than the hackers, but my cooperation changes my focus. If I have been complicit, it is *my* behavior playing on loop, not theirs. Even if I get pickpocketed or my car is broken into, I may have regrets about where I parked or whether I forgot to zip my purse, but those regrets land differently, more diffuse, less entangled in self-blame. If I agree to a scam, it's like I taped the Kick Me sign on my own back. That is the essence of the sucker's regret.

What purpose does it serve us, individually or as a species, to feel bad about playing the fool? The most basic answer is that a bad feeling like regret promotes learning. As a human parent with two children, I am often grateful that we find the experience of regret so deeply unpleasant. Regret is an effective teacher for a kid tempted to put his hand back on the hot stove or pull the cat's tail again and see what happens this time. There is learning about scams, too. Unlike other kinds of overt threats—fire, or lions—it takes some experience to figure out which kinds of situations are traps. By definition, con games do not announce themselves. Learning how to spot them yields benefits. People are regretful of outcomes that they could have controlled; you want to learn to avoid the bad outcomes you can avoid. By contrast, fear of things you have no control over is not very productive, because it yields no action items and has no natural boundaries. It makes sense to think you should spend more time mulling your investment losses from a Ponzi scheme than from an unforeseeable market drop, because dwelling on the former serves an educational function that dwelling on the latter does not.

The value of regret is that it leads to learning—but most of us know that regret can be a mean, zealous teacher, insisting on lessons we will overlearn and warnings we will over-heed. People develop exquisitely sensitive radar for mistakes that will yield footage for the cringe reel, and the biggest red flag is situations soliciting cooperation. In Ralph's Garage and in "Free Money!" the subjects saw setups that by their very structure previewed the self-recrimination to come. Both studies asked people to make a choice, to affirmatively engage in an optional transaction. That alone would be enough for subjects to preview exactly how it might go wrong: bait, switch, regret.

The difference between a scam and a robbery is that the mark appears to part with his money voluntarily, whereas the robbery victim is not complicit. Sometimes the voluntariness is even less salient but rather merely any kind of active cooperation with a scam. Imagine for a moment that I belong to a group where members have to pay dues. After years of faithfully paying my dues, I learn that others, also members in good standing, routinely underpay. Indeed, imagine being an American taxpayer and learning that while I report honestly on my income, many of those wealthier than I am do not. There are a lot of plausible reactions to this scenario: frustration with failures of accounting or condemnation of my greedier peers. Not least among these responses is the cascade of emotions that might flow from my sense that I've been a chump, a literal loser.

Regret is a workhorse of an emotion, because it sucks in real time and also looms overlarge in the hypothetical. We experience regret all the time, of course, and our deep experience with that emotion makes us feel like we understand it, but there is also a puzzle at its heart: regret sometimes comes untethered from the harm that unleashed it and goes on to attain an emotional status of its own. When "George Williams" or eager undergraduates try to flag down their next study participant, the targets are already mentally playing out the consequences of agreeing. What could go wrong if I agree to this? What could go wrong if I ignore it? One of the implications of the sucker setup is that the emotional fallout looms large remarkably independent of the material harm. In both of the studies, the practical downsides of cooperation were relatively limited, but I think they still elicited that feeling of anticipated regret.

Imagine that I see a snake sunning itself on a rock and decide to reach over and pet it. The snake bites my hand and it hurts. My feelings of regret for my choice would hopefully cause me to refrain from snake petting the next time. That's a great use of regret—thanks, evolution—and as an empirical matter, we know that people do in fact try to avoid the snake or the hot stove the next time. The twist is that people do more than that. They try to avoid the harm, of course, but almost on a separate cognitive track they try to figure out how to avoid the feeling of regret. Feeling regretful stings by itself, even independent of the pain of the bite.

If you think about regret for a minute, you'll probably guess that you regret bad choices, and the amount of regret you feel is roughly proportionate to the badness of the choice. For example, I should regret gambling away $100 more than I regret gambling away $10. This is in fact a good, rough estimate, but in real life there is a predictable caveat. It turns out that what triggers an acute regret response is not necessarily that something was really bad. For example, if I park my car in its normal spot and lock it as usual and it gets broken into overnight, I feel anger or sadness but not much regret. What could I have done? If I don't buy a raffle ticket and someone else wins, I feel fine: I'll never know if the winning ticket was the exact one I passed up.

What triggers regret is when you *know* you could have had something better and you *see* how you could have gotten it. Some phenomena trigger that perseverating what-if self-recrimination and others do not. In sucker's games, the regret stakes tend to be lopsided. The fear of feeling regretful, regret aversion, causes us to materially change our behavior by favoring decisions that seem less likely to produce regret later on. This is how I have been talked into buying rental car insurance. ("But, ma'am, how are you going to *feel* if you damage this car and know you could have had it paid for?") What produces regret is the *knowledge* of what might have been, not the poor choice per se.

Go back to the "Free Money!" example for a minute. There are two ways that the interaction might end in regret. One is that you walk over to the table and it's a scam; they want your Social Security number and your email address in return for that dollar. Oops, bad choice. The other is that you do *not* walk over to the table and the money is in fact totally free and available. Hm, also bad, but not the same. Both of these are unwanted outcomes, but the regret implications are not equivalent. As soon as you spot the table, you know that only one of your choices will risk true regret. If you walk by the table and in fact it is legit, you will probably never be the wiser—just like the person who hangs up on the stranded driver will never know how the story ends. There may be some nagging doubts that you made the wrong choice, but not that certain verdict. If you walk over to the table and try to accept their offer, though, you will learn for sure if you're the fool or not.

The inexorable connection between regret aversion and overblown sucker fears feeds distorted decision-making, leading to choices that would seem to be in conflict with deeper preferences and values. For example, imagine that my cousin calls me to ask if she can borrow $500. She promises to pay me back as soon as she gets her next paycheck. Let's say that I have the money, and I think the right thing to do is to help her out—I *want* to help!—but I can only afford it if she's really going to pay me back.

There are two mistakes I am trying not to make: mistaken trust (lend and don't get repaid) and mistaken distrust (don't lend when I would have gotten repaid). Morally speaking, I would rather mistakenly lend than mistakenly withhold. But regret aversion research suggests I'll err in the other direction, because regret takes hold most viciously when I'm certain I made a mistake. People predict which kind of scenarios are most likely to yield an ex post facto feeling of regret and then avoid them. If I lend, I will have unequivocal evidence of whether I was duped, because either she pays or she doesn't; I'm setting myself up for the possibility of regret. If I don't lend, I'll probably never know what she would have done, because there's no step two to our transaction. I say no, she figures something else out, and I probably don't even hear about it. I might feel some misgivings, but I won't have that in-your-face evidence that I made the wrong choice.

When people see a cooperative ask—Can I get a loan? Do you want to be my partner, or my investor?—it sets off their regret receptors. They get the input *I'm being asked to trust someone* and that cognition connects automatically with the thought *If this goes badly, the regret is going to feel terrible*. And psychological research has shown that it is not just the loss; it is the advance fear of misplaced trust. In a simple study on regret, psychologists asked subjects to imagine they had $100 to invest. They were told: *Invest in this company and there is an 80 percent chance you get your money back (no more no less), a 15 percent chance your money doubles, and a 5 percent worst-case chance you lose it all.* One group of participants was told that 5 percent downside risk was a risk that the founders were actually fraudsters. The rest were told that the downside risk was just that they had overestimated consumer demand. Participants who were willing to invest around $60 with uncertain consumer demand were only willing to invest $37 if the same level of risk implicated misplaced trust rather than market forces. Investors who were risking personal betrayal demanded a discount.

The Tragedy of the Commons

Whether I decide to lend money or shy away from one investment opportunity, these are decisions that speak to a complex personal calculus. But, crucially, these are also the kinds of decisions that implicate social cooperation. In turn, the fear of playing the sucker can have real consequences not just for individuals but for relationships, communities, and society. Who trusts whom in what structural contexts is the definition of a social order.

Social cooperation is a sucker's thicket. That proposition got the formal theory treatment in 1833, when the British economist William Forster Lloyd wrote a famous essay on unregulated livestock grazing on the common pasture that introduced the concept that became known as the "Tragedy of the Commons." The Tragedy of the Commons is a fable and a math problem at once. As a public commodity, the pasture on a commons is only sustainable if its use is rationed, because once it's overgrazed, it's useless to everyone. This simple setup makes for a perverse hierarchy of incentives for the individuals vying for access. The commons works best as a resource when everyone derives a moderate benefit. Any one family with a couple of sheep can sneak an extra sheep or two out there without ruining the whole commons, but if too many families try it, the grass dies and the resource is destroyed. It is a perfect sucker's dilemma: Take advantage, or risk being taken advantage of?

Economists have theoretical predictions about what happens in a situation like this, and now we also have in-person data for what really happens via experimental games. The Tragedy of the Commons has often been studied with a setup called the "Public Goods Game." The origin of the Public Goods Game, like most economic games, is the question about how people navigate between the selfish best-case and the communal best-case scenarios.

Experimental games pare human transactions down to the very barest bones, to make it as easy as possible to infer people's preferences and intentions from their overt behavior. Games like this come up a lot in the study of sucker's dilemmas, and they have some consistent features. The research subjects, often students but sometimes people from the community, are given a set of instructions for how some series of transactions is going to go—i.e., the rules to the game. The players normally do not see one another or at least they do not know who is in their group; communication is via anonymous envelope or computer interface. The players know that all the money they are trading is real and that they are really playing with another research subject, not just an experimenter pretending to be a player making spontaneous choices. One of the key features of this kind of research is that there is no experimenter deception.

In the Public Goods Game, players are organized into groups of four. In each group, every player receives ten $1 bills in an envelope (this is often called the "endowment"), and they are told that they have a choice to make.* Each player can put any amount between

---

* Different versions of a game use different procedures for the nuts and bolts of how players exchange money, how they indicate their contributions, and how much money they get to start. For clarity of explication, I use the traditional $10 figure and describe a physical exchange procedure (money in an envelope), although especially for contemporary studies this is almost always done via interactive computer program.

$0 and $10 into the "pot," the temporary communal account that gets shared at the end. After all the contributions are made, the total amount in the pot is multiplied by one and a half—so, by design, the whole is greater than the sum of its parts. If all the players put in $10, for example, the $40 pot would grow to $60. Since the pot gets divided evenly at the end, a $60 pot means each player would go home with $15. If each player put in $5, they would go home with $12.50 apiece. (The $20 in contributions becomes $30, meaning $7.50 per person from the pot plus the $5 in reserve.) The "public good" here is represented by that multiplier, in the sense that everyone can have access to something if there is enough cooperation to create the surplus.

To put this in the language of the Tragedy of the Commons, if everyone contributes their full endowment, that's full cooperation. That scenario mimics the world in which each person gets to graze as many animals as possible right up to the threshold of overgrazing. The commons is in top possible condition. If each person is only moderately cooperative (contributing $5, for example), that's like a situation where they may get a little extra grazing, but they will also notice that the quality of the grass has worsened. If everyone overgrazes—i.e., no one shares—they are all forced back onto private land; the commons is dead and there is no public good.

What would you choose? What would you be hoping for, and what would you be afraid of?

Let's say that I'm basically a free-rider-personality type: I want to drive a Hummer and run year-round air-conditioning but live on a clean planet thanks to everyone else's abstemiousness. If I am playing a Public Goods Game, my hope is that everyone donates—except me. Each of the three other players contribute $10 each, and I hold mine back. The communal pot would then have $30 in contributions, so $45 to divide. Every player takes home one-quarter of the communal pot, or $11.25.

But of course I did not contribute, so I get that $11.25 *on top of my original $10*. I leave the game with $21.25! I get $10 more than any other player and more than twice what I started with. The person who doesn't cooperate while everyone else does play nice comes out on top. In game theory language, this selfish behavior is called "defecting."

And as a matter of strategic theory, no matter what the other players do, I am better off defecting. If the other players cooperate, I can go home with the fruits of their cooperation plus the seed money that I declined to contribute. If the other players do not cooperate, I am *also* better off not cooperating. After all, if they are going to keep their $10, I shouldn't throw mine in to divide among them.

Unfortunately, if that is true for me, it is true for everyone else, too, meaning that the equilibrium solution—the theoretical prediction of what will happen if everyone is rational—is pretty bleak. The game starts, and everyone gets $10. No one contributes any to the pot. The game is over! There is a reason they call economics the "dismal science."

When economists and psychologists started running this game with real players in the mid-1970s, one thing became clear right away: most people do contribute to the pot, and most contributions are substantial. Not so dismal after all! A lot of people choose trust and they try to create some public good. But most players do keep some money back, and if players play together over multiple rounds, the donations will get lower and lower with each new round. Researchers wanted to know: When people withhold cooperation, is it because they are they greedy, or because they are scared?

Economists predicted that greed would be a primary motivator; they thought people would go for the highest possible payoff and hope others would be more naïve. But what they found over time caused some reconsideration. Their subjects were actually quite interested in cooperating—but worried about looking foolish. Robyn Dawes, Jeanne McTavish, and Harriet Shaklee, researchers at the University of Oregon, devised a clever test to distinguish between the selfish and the sugrophobic. They asked each player two questions: (1) what are you going to choose, and (2) what do you think everyone is else going to choose? Each player wrote down their answers and turned in their predictions.

What Dawes, McTavish, and Shaklee worked out was that the logic of sugrophobia is different from the logic of selfishness when you're playing with people you think are going to contribute. If you're a greedy player and you find yourself in a game where you expect everyone else to contribute, that's your chance to go in for the kill. If greed was motivating the players, the researchers should see that those who predicted cooperation from others would *decrease* their own contributions. That is not what they saw in the data. Instead, they found that players who expected to be treated cooperatively would *increase* their own contributions. When the subjects were less scared of being a sucker, they were more likely to cooperate.

The research team noted an additional and unexpected observation: the players were frantic, and furious, when others appeared to have played them for a fool. They were not impassive rational actors making their bets and taking their licks; they treated stingy players as "double-crossers" or cheats. In their bemused write-up, Dawes, McTavish, and Shaklee reported:

> It is the extreme seriousness with which the subjects take the problems. Comments such as, "If you defect on the rest of us, you're going to have to live with it for the rest of your life," were not at all uncommon. Nor was it unusual for people to wish to leave the experimental building by the back door, to claim that they did not wish to see the "sons of bitches" who doublecrossed them, to become angry at other subjects, or to become tearful. For example, one subject just assumed that everyone would cooperate . . . and she ended up losing $8.00 which matched the amount of money her friends had won. She was extremely upset, wishing to see neither the other members of the decision group, nor her friends.

One cooperator even yelled at his defectors, "You have no idea how much you alienate me!" Indeed, the researchers had planned to try versions of the game in which players could interact with one another face-to-face, but seeing how high the feelings ran, they desisted: it would not be ethical, they thought, to make people this miserable with one another for an economics experiment.

Much like regular games—Monopoly, or The Settlers of Catan—setups like the Public Goods Game offer a model of a mini-society. People cooperate and transact and experience the joint and individual consequences of their choices. What researchers are sometimes surprised by is that their subjects start to really live in the metaphor. (For anyone whose sibling or cousin has ever over-turned a Monopoly board in fury, this presumably resonates.) The succinct elegance of the Public Goods Game is that it shows how each individual cooperative calculation added up, for the players, to a social order. Players who went home with small payouts were not frantic about the dollars they missed; they were frantic about the alienation.

At a deep instinctual level, the sucker construct is not about material payoffs or outcomes; it is about social standing and re-spect. The game's cooperators, the fools, felt like social pariahs. Even in a fake, irrelevant simulacrum of social life, the pain of ostracism or status demotion was intense. Alain de Botton, who coined the now-familiar term "status anxiety" in his book of the same name, wrote about status as a kind of love: "To be shown love is to feel ourselves the object of concern: our presence is noted, our name is registered, our views are listened to, our fail-ings are treated with indulgence and our needs are ministered to." Most people care if their society loves them, he wrote, and that helps explain why "from an emotional point of view no less than a material one, we are anxious about the place we occupy in the world." Even at the level of automatic cognitive processing, peo-ple perceived to be higher status attract more "visual attention"— meaning other people are more likely to look at them and to follow their gaze as well. People find them more interesting, and easier to remember·

Most of us know where we stand, globally or in specific do-mains, and most of us care. My own stock changes dramatically

if I walk from a law school classroom (all-time high!) to my son's
high school cafeteria (so low that he would absolutely never let
this come to pass). It is tempting to be dismissive of status, because
it seems too nebulous and too shallow to matter. Indeed, we have
derogatory language for people who openly seek status—they are
"status conscious" or "clout chasing." An acquaintance recently
noticed a ficus tree in my living room and informed me that it was
the "status plant of the year." I don't think it was a compliment. To
care about status is to be vain or superficial, or at least that is how
we talk about it.

In fact people care deeply. In some ways it is easier to grasp the
stakes of status by thinking about it in the negative rather than the
positive. What status bump do I get out of having an indoor tree or
a nice car? Even the question feels trivial. But I know that if I were
moved into a smaller office at work, or not invited to a contracts
conference, I would not find it so funny. Social demotion is humiliating, like being broken up with but by a lot of people at once.
And a crucial fact for any sucker analysis is that being made a fool
is always a demotion.

Indeed, if regret is aversive, humiliation is downright horrifying, which is why there is a satisfying literalness to its translation
for the horror genre. I remember when I first saw the movie *Carrie* in high school, oddly enough for a film class. I was a senior
and had a little bit of emerging perspective on the social life of
teenagers, not to mention a very specific local perspective on adolescence in rural Maine. Stephen King clearly grasped the lurid
terror of the sucker's humiliation, and the fertile ground of high
school, when he wrote it. A lot of storytelling (movies, TV, literature) features humiliation, usually in the form of laughing and
pointing, but King, and later Sissy Spacek, got to the dark heart
of the terror. Carrie, as you may know, was a social misfit navigating a complex adolescence (abuse, telekinesis) when she was
invited to the prom by an unlikely suitor, the popular boyfriend of
a classmate. She was suspicious but she agreed; she and her date
had a nice time, and it turned out that the invitation she feared

was a scam was in earnest. Unfortunately, this set her up to be vulnerable for the real betrayal, when she and her date were voted prom king and queen. Lured to the stage with the promise of coronation, she was instead positioned below a bucket of pig's blood. Cruel classmates tipped the bucket, soaking her dress, reminding everyone of an earlier menstrual disgrace and making clear that her enjoyable night of social acceptance was ultimately a ruse. In falling for the scam, she was made literally contaminated, physically unfit to remain in society.

Sometimes social rejection, as a form of pain, can seem from the outside as if it were not real pain but more lyrical or metaphorical—but that's not how it feels. Humiliation is quasi-violence, and it's experienced by its subject as such.

This appears to be true even at the level of neurological processing. Two Dutch social psychologists fitted their experimental subjects with EEG caps in 2013 (think electrodes on the skull) to measure neurophysiological activity. They told the subjects various stories with different emotional profiles. They found that, compared to stories that elicited happiness or sadness or even anger, humiliation was the most intense emotional experience, a brutal layering of self-blame on top of rage.

"Humiliation is a personal experience," wrote other (also Dutch) researchers, "often stemming at least in part from a sense of inferiority." The public humiliation of being duped is not just an act of active status renegotiation. It is a reminder of who holds the cards, a ratification of the existing hierarchy. In high school terms, bullies don't target the captain of the football team. Being targeted is a reminder of vulnerability; if you were that elite, no one would have tried to sell you the Brooklyn Bridge in the first place.

❖

The sucker dynamic is so potent that it feels terrible even when it has no visible consequences at all. When I was in high school, there was a prank, deployed with a wink but also a nod to the cruelty in the holster. The joke was that a boy would approach a girl—say, at a noisy party—and lean in to ask, "Would you like to dance?" Whatever the girl's response, he would respond loudly, "No, I said you look FAT in those PANTS." A perfect sucker's morsel! Two people engage in a social exchange that purports to be about an invitation to dance and turns out to be an insult, a deception, and a rejection. The genius of this joke was that it also worked if she declined—what matters is that she entertained the compliment, whether she rejected it or not.

The fact that there are no material stakes gets to the heart of the issue: a sucker's game is a power play, at least in the microcosm of the particular interaction. Even if no one is asking for money or promising the Brooklyn Bridge, there is still a transaction, there is still something at stake, there are still winners and losers. What is being transacted is social power. Anyone can lob an insult, but the insincere compliment doubles the speaker's power; it makes the target both a victim and a sucker. It asks the girl to cooperate in the interaction—even, best-case scenario, to evince pleasure at being in the speaker's good graces. It doesn't always work, of course. Some bullies are ostracized for their cruelty or overestimate its efficacy for social power. But the goal is the one-two punch of the insult—there's a reason it's called a "put-down"—*and* the temporary status boost of putting one over.

The pain of thinking you belong when you do not is something that adolescents understand with the deep knowledge of people who can only trade in social capital. (This is also true of academics, who, like teenagers, have limited access to material wealth, ample sites of status contestation, and a lunchroom.) The joke pares the sucker construct down to its nub, because the only transaction is a status exchange. The only bait is the bait of human connection—you're in!—and the only switch is revocation—haha, you're out! Social respect is at the core of how we navigate sucker dilemmas all the time. In a sucker's game, status makes the rules, and status is the prize. Do you think *I* would want to dance with *you*?

La Nausée

The cooperative undergrad in the Public Goods Game was right: the force of being suckered lies in the gut punch of alienation.

Sucker fears are hard to articulate, and they don't feel good to acknowledge. As humans, we are averse to feeling regret and deeply, profoundly scared of social derogation. Ostensibly egalitarian societies like ours are rarely frank about social status and hierarchy, forcing our questions into innuendo and half jokes, but the accusation that you have played the fool is dead serious. Most of us are not going to get blood dumped on our heads; we are just going to feel insulted. Nobody wants to be sensitive to the low-stakes mockery of a possible phone prank, but most people are.

There are some stimuli in this life that have outsized power. In 1951 the psychology graduate student John Garcia left his graduate program to work in an animal lab at the U.S. Naval Radiological Defense Laboratory in California. As part of one of his early studies there, he exposed rats to sweetened water at the same time as radiation, a coincidental artifact of the experimental design. Rats love sweet tastes, but radiation is terribly nauseating. The "Garcia rats" refused sweetness of any kind for weeks after.

If you have ever had food poisoning, or a terrible hangover, you may sympathize. Psychologists even coined a phrase, the "Sauce-Bearnaise effect," to describe the intense aversion that comes when nausea is temporally but coincidentally associated with the experience of a new taste. Nausea turns out to be an efficient inducer of overlearning: hungover, I'm not just avoiding Jägermeister; I'm revolted by both alcohol and licorice, and for a long time. Social denigration works the same way. Humiliation and self-recrimination feel awful. The brain races to explain what happened. It builds out a cognitive apparatus to patrol for the threat next time. No sweets, no liquor, no cooperation. Not hungry, just suspicious.

CHAPTER 2

Weaponization

I grew up off a dirt road in a small town in southern Maine. It was quite rural but not as remote as it sounds, and actually most of the roads around us were asphalt. Nonetheless, our road stayed unpaved, due in part to a shared local view that when the town paves, they get in your business. Mainers are temperamentally distrustful—individualistic and deeply skeptical of elites and institutions, especially the government. They will respond to a story about unscrupulous dealings or untrustworthy summer people with a headshake and a "That's how they getcha."

I had this suspicious prototype in mind when I came across George Fournier, a man from Aroostook County whose bungled estate plan showed up in a casebook when I was teaching a course on wills and trusts. Aroostook County is the largest and northernmost county in Maine, and in-state it is referred to without the proper noun, as in "The new guy at work moved down from the County." Pre-cable and pre-YouTube, they were rumored to have a whole television channel devoted to potatoes up there.

In a region with a median annual household income of about $30,000, George Fournier, a "very frugal" bachelor, accumulated hundreds of thousands of dollars, which he kept hidden around his house. No one, including his attorney and his beneficiaries, knew how he had amassed his wealth. Toward the end of his life, he packed two boxes with $200,000 cash each and delivered them to his neighbors, giving them verbal instructions for the boxes when he died. When the time came, though, the neighbors proved to be confused about what they were supposed to do with the money, and the whole thing wound up in court—an ironic outcome, because George, by all accounts of sound mind and canny financial prowess, had gone out of his way to avoid banking and legal intermediaries. He appeared to distrust financial institutions, preferring to keep his assets liquid and in arm's reach. He certainly did not trust the government, whom he hoped to keep out of his affairs at probate by writing nothing of his elaborate scheme down on paper. He had not even told his own attorney about the stashed boxes of cash. George walled himself into a fortress of suspicion.

Like most people, I am on a spectrum somewhere in between the unbanked hermits and the blissful rubes. I understand what George Fournier was worried about, and although I come down in a different place, I respect the set of concerns he had as serious and deep. I am more credulous, or maybe more resigned, than he was. I use Google and Twitter, though not Facebook. I trust my retirement savings to Vanguard—index funds only, though, because I'm skeptical of fund managers. Personally, I don't think my employer is putting one over on me, but I'm not sure about my senators. I don't think it's unreasonable to fear the Man (out to get you), or Big Brother (always watching), the Washington elites, the Mainstream Media, or the Dark Money.

Versions of exploitation narratives transcend class and party. Fox News viewers may suspect subterfuge from liberal elites, Anthony Fauci, and George Soros, while Biden voters suspect Fox News itself of being a horrifying grift. These are explicit, familiar debates about exploitation and power: who can be trusted, who is owed, who has been promised more.

The question of who is using their power duplicitously *is* what we are talking about when we debate politics or civic life—Karl Marx theorized a revolutionary political movement on the proposition that capitalism is exploitative by definition—and I think most people can speak openly about parts of their lives in which they themselves feel exploited by those in power. The narrative of exploitation can offer a framework for understanding a shared experience. There is value to the vernacular shorthand for power grabs: when we talk about "Big Pharma" or "Big Tobacco," we are warning each other of real threats. It works in popular culture, too: Think of the rallying cry from the unlikely poet laureate of office life, Dolly Parton, who reminded ambitious 9-to-5ers that they're "just a step on the bossman's ladder." It's a catchy indictment of the American social order as the promise of mobility and the practice of dominance.

To make an extraordinarily banal statement, when the powerful exploit the weak, the political and material stakes are very high. Whose labor is rewarded, and who works for free? Who makes choices, and who pays for consequences? The answers to these questions are at the root of the hierarchies that allocate wealth and power. The discourse of exploitation—the creation of shared knowledge of how the strong manipulate the weak—is an important social and political tool. Rhetorically, it is a discourse that can motivate social change or even rebellion against the powerful. That's how they get you! Wake up, sheeple!

And yet, at the level of psychological threat, the prospect of top-down exploitation packs a surprisingly feeble punch. I think many people feel pushed around by the government but not truly suckered, or exploited at work but not quite duped. Our explicit civic value system tells us these top-down harms are serious, because breaching the trust of the vulnerable is a serious abuse. Nonetheless, what sets off the five-alarm psychological sucker panic tends to look different. The real sugrophobic triggers are implicit threats that interlopers, hustlers, and people on the lower rungs are going to climb over you if you let them.

What makes sugrophobia a little puzzling and a little perverse is that it is not exactly a fear of the exploiter; it is a fear of being the fool. It's a fear not of the aggressor but of what the aggression makes *you*. The distinction implies a crucial shift for what experiences will set off the phobic response. For reasons of habit or status quo, it does not feel shockingly insulting to see the self as the pawn of the government or the rich. But if I can be exploited by my peers or (worse?) my subordinates, what does that make me?

The patriarchy and the aristocracy, even the meritocracy, are arguably forms of structural exploitation, the hoarding of power at least partly by deceit. But sugrophobia is aimed squarely in the other direction—think welfare fraud or immigration scams—at the weak making fools of the powerful. It is a more shadowy deployment of the exploitation narrative. One way to keep people subordinated is to tell stories about their scheming intentions, to covertly leverage the fear of duplicity as a weapon of manipulation.

Indeed, the weaponization of sucker rhetoric is at the root of recent political experiments in sugrophobic populism. Starting in 2011, Donald Trump began to lay out what became his signature theory: that white American men are getting scammed. From the beginning, he planted his political ambitions in the fertile ground of racial resentment. Ta-Nehisi Coates described him in his 2017 piece for the *Atlantic* as "The First White President," with a political career that originated "in advocacy of birtherism, that modern recasting of the old American precept that black people are not fit to be citizens of the country they built." Indeed, birtherism was a movement built around the claim that Barack Obama, a successful, credentialed, charismatic politician, was putting one over on his constituents. Coates observed a demagoguery rooted in the furious insistence that Barack Obama was an impostor:

> After his cabal of conspiracy theorists forced Barack Obama to present his birth certificate, Trump demanded the president's college grades (offering $5 million in exchange for them), insisting that Obama was not intelligent enough to have gone to an Ivy League school and that his acclaimed memoir, *Dreams from My Father*, had been ghostwritten by a white man, Bill Ayers.

Trump took what he understood as a widely shared fear of being made a sucker and found its most pernicious pressure points. The effects of his outrageous, defamatory claims were powerful. It was not just that President Obama was unqualified for office; it was that the president, a Black man, had made a fool of white people by lying that he was American and pretending to be smart. Indeed, the visceral, frantic aversion to being a sucker (accused of being manipulated by Russia: "No puppet. No puppet. You're the puppet") was one of the only consistent political convictions of the Trump presidency. Trump knew, presumably from personal experience, how powerful a motivating force he could unleash by insinuating that his supporters—the white working class in particular—might be taken for a ride by a Black upstart.

The political career of Donald Trump offers a cascade of evidence that the sucker frame is a weapon in the right hands. All it takes is a little nudge, and suddenly paying your taxes feels like cooperating in a sham. Trump laid bare the fragility of moral norms, their susceptibility to the bully's accusation that, actually, doing the right thing is for chumps. A snide remark (on Mexico: "They're not sending their best") makes humanitarian asylum look like a reward for cheating the system. Indeed, as the journalist Anne Applebaum wrote, the Trump administration took the counterintuitive but effective position that "morality is for losers." His obsessive policing of illusory scams and hustles led to children in cages and trade wars with China and was no small part of his political appeal. He pressed on the shameful fear like pressing on a bruise.

Trump's success was built on more than just distrust of government; it was built on a warning that your (weak, bleeding-heart) government would make *you*—the default "you" is always the white man—silly, womanly, defeated. Obama is a hustler, and he's made you the fool, argued Trumpism. (Coates made this connection vividly in his essay, describing the sexual innuendo of the Steve Bannon philosophy of domination: that white men made themselves "cucks" when they submitted to Obama's presidency, as they would if they submitted to Black male conquest over their wives.)

Birtherism was intended to invoke the prospect of white men as suckers in a way that warnings about super PACs never could. Every aspersion cast on Obama was a warning to those on the fence.

What Trump understood was that the scariest hustlers are the strivers. The fear of being a sucker is at root a fear about disruption of a status hierarchy, and that hierarchy is most grievously disrupted when the scam comes from peers and subordinates. This theme, though more subtle than the boss man tropes, wends its way through fables of con artists and hustlers.

Who are the archetypal grifters? They are low-status people lying to part the higher-status fool from his money. They are Black people tricking white people or women scamming men: Will Smith in *Six Degrees of Separation* or Elizabeth Holmes and her Silicon Valley investors. Gold diggers, hustlers, and con artists— these are the marginalized fringes, duping people of means to get access to their wealth. When people with power are exploitative, it is infuriating, frustrating, demanding of a call to arms, but it is also, at some level, business as usual.

If you think about classic cons, they are always stories of clout chasers and interlopers—impostors pretending their way up the social ladder or into the inner circle, Holly Golightly or Tom Ripley characters, pretending to shop at Tiffany's or graduate from Princeton to get a leg up. Charles Ponzi himself was a striver, an Italian immigrant who was working as a mining camp nurse, fresh out of prison, before he developed his schemes. (Bernie Madoff was more of an insider from the beginning, which perhaps helps explain his decades-long success: he didn't set off the radar.)

The Sucker Schema

Why does our radar for fool's games go off at some times and not at others? Sometimes someone will openly pull the trigger, naming a dynamic in explicit terms. Trump, for example, has repeatedly invoked suckers in improbable contexts by specifically ridiculing war heroes as losers or calling asylum seekers con artists. In other situations, one doesn't necessarily use the verbatim language of fools, but the threat of a scam is clear, because we've learned from past experiences the risk we are taking. If I am buying a car off a used-car lot, I am on guard; I know that the salesperson's goal is to get me to part with more money than I need to. If I am selling something illegal, I know that I should be on the lookout for a double cross, because I get the incentive structure and I've seen a lot of heist movies. But most interactions are ambiguous, more open to interpretation. Birtherism managed to take a straightforward fact—a Black man born in Hawaii was running for president—and make it sinister. The demands for paperwork and proof accompanied a steady stream of innuendo that Obama was lying about his nationality and even his religion, perpetuating longstanding racist claims that Black Americans are more foreign and less American.

The threat of a sucker's game looms large even if it's just a nagging feeling at the back of your mind or a sudden shift in focus that never gets a name, and it can be hard to foresee what will raise your sucker hackles. The interpretation of any experience depends on some cognitive science about how we process information about the world. To zoom out for a minute: at a general level, the way we figure out what is going on—*What am I looking at? What is expected of me? What happened here?*—is to match perceptual stimuli (sensory information) to mental models (concepts). Just as we have a mental model of a bird or a car, we also have a mental model of a sucker's game. Psychologists refer to a mental model as a schema. It is a framework for organizing and understanding information related to some phenomenon. We see things, hear things, feel things, and then use existing mental models to draw conclusions or make judgments.

For example, when I see an animal and want to figure out if it's a bird (most of this would be happening automatically, not at the level of conscious processing), I digest some information in terms of the bird schema. If an animal has feathers and a beak, it is likely activating a whole mental structure. Now I know to ask if it flies and lays eggs. I don't inventory every feature of the animal individually every time; I see the feathers and the beak and now it's just a matter of what kind of bird it is.

What's remarkable about this mental organization is that there are real neural networks at play. Once I have thought of feathers, I am literally quicker, at the level of electric signals among neurons, to think of eggs and flying, and slower to think of fur or scales. People who read about eggs and are asked to fill in the blank on the word _EATHER will be more likely to put in an F than an L or a W. If I see the bird move, I am more likely to perceive flying than walking, even if the visual stimuli are open to interpretation. The schema does a lot of work to organize and simplify a complicated world.

The same psychological predictions and patterns apply whether the schema is describing something relatively concrete and settled, like the category of things called "birds" or "eggs," or something intangible and complex like geometry, or friendship. When people think they are seeing a con game, they are quicker to see deception, or ill motives, or contempt in otherwise ambiguous data points.

We have schemas, also called scripts, for a lot of human relationships. I have a mental model that tells me what to expect when I sit down at a restaurant and am approached by a woman wearing an apron and carrying a pad of paper, and that guides my behavior in turn. I don't tell her about my relationship with my mother and I don't ask her for a ride downtown. I don't even offer her money, even though I plan to pay for my food; I know that the script goes: (1) order; (2) eat; (3) pay. Just as I have a schema for "ordering food," I have a schema for "playing the fool." There is a sucker schema with consistent features, and the activation of the schema has real consequences. To control the schema's activation—framing the situation, controlling the narrative, or just making the sucker accusation—is to wield a powerful weapon.

When my kids were very young, we went on a family trip to St. Louis to visit my sister. At the time, my daughter was just two years old and my son would have been almost six. We live in the city in Philadelphia, not so dissimilar from my sister's neighborhood in St. Louis, both dense residential blocks adjacent to the downtown. However, we live in a row house, with two party walls and a stoop on the sidewalk. In St. Louis, the houses were close together but had narrow side driveways and mowed front yards. For whatever reason of norms or zoning, those front lawns were rarely fenced from the street. As we walked along, my son, impressed, remarked, "Everyone here has their own park." My daughter tried to toddle from lawn to lawn.

My kids saw flat mowed grass, adjacent to a city sidewalk, and activated the "park" schema. My son was old enough to perceive some ambiguity in the norms of these individual green spaces, but my daughter was thrilled. The activation of the park concept affected her behavior and her expectations: she predicted relaxed rules (yes to running, no to hand-holding) and leisure. And, of course, "park" is a happy thought.

When my husband and I saw the grass, we thought "other people's yards": they looked nice enough, but we were not very excited and we did not try to start a game of tag. You see some stimuli—grass, city—and you infer what it means. The meaning that gets activated guides your behavior and your feelings. The stimuli themselves were open to more than one interpretation, and we each behaved in line with what we thought we were seeing.

If it's possible to disagree about the meaning of grass, it's downright unavoidable for complex social interactions. Social dynamics don't define themselves, so sometimes you see a sucker's game because of some little reminder, a little red flag that cues the fear; other times the same basic dynamic just doesn't push those same buttons. Whether or not we perceive the situation to be a sucker's game is contingent, and it's also consequential. If the sucker schema does get invoked, it matters, because it has profound psychological and behavioral effects.

When people hear about George C. Parker, they see the sucker's game right away. In the late 1800s, police had to remove several prototypical rubes who were attempting to set up tollbooths on the Brooklyn Bridge. They claimed that they were bona fide purchasers, having bought it—many for hundreds of hard-earned dollars—from Parker, a person who did not, of course, have the legal right to sell it. The outlines of this story—the slick salesman with a bald-faced lie, the greenhorn blithely volunteering for humiliation—are the apotheosis of the sucker. But much of the sugrophobia that we care about—in gender, in race, in politics, in love—will never look like the prototype, even though the sucker radar is going off in our brains. The stimulus itself is not decisive; it is open to interpretation, and that interpretation depends on our experiences, our habits of mind, and the situational cues telling us where to focus our attention.

To see how this might play out in an everyday context, think of a routine request at work. In my case, I work as a professor, so I'll draw from that. Imagine that a student emails me and tells me that she has experienced a death in the family. She is writing to inform me that she will miss class and would like to request an extension on her final paper. My default instinct would be to think about the appropriate etiquette of condolences. In light of a student's description of a loss in the family, I implicitly ask myself what my role is; I surmise that my job is to express sympathy, change the deadline, and refer her to the dean of students' office in case she needs more support services.

Now imagine a rewind. I am about to meet my student and I run the situation by a colleague, who smirks, "That's quite a convenient coincidence! If you believe that, I've got a bridge to sell you." As a friendly female professor in a historically formal and male-dominated profession, my hackles are now up. I do not want to be seen as weak.

The sucker schema has been triggered, and now I am braced for my meeting with a new mental framework. I ask myself what my job is and I see that, as a responsible faculty member, I need to make sure students are complying with the honor code. After all, this class is graded on a curve, and it's unfair to everyone if some students are gaming the system. Still pleasant, of course, I ask the student to submit documentation to verify the emergency while I look up the formal school policy on which family situations may justify accommodations. In the shuffle, I forget to offer my condolences to her family. I leave the conversation hoping I'm not being duped; she leaves wondering if the extension was worth the alienation.

Invoking the sucker framework redirects a whole interaction away from one set of interpersonal protocols—the protocols of condolence—to another: the protocols of documentation. Whether I notice it or not, now both my student and I feel disrespected and alienated.

Human interactions are subject to a wide variety of interpretations, and unless the shift in interpretive frame is flagged as such, we probably don't even think about the counterfactual: How would I act if I did not care about being duped? What is more important to me, being kind or being savvy? The sucker threat disconnects people, making us feel unseen in the eyes of our fellow beholders. My colleague's snide remark is an explicit cue; he is telling me in so many words, *I think you are being scammed.* And when he does this, he is changing how my student and I experience our relationship. (He is also changing our own dynamic by insinuating that he is savvy and I am gullible.) I am worried that the student is trying to disrespect me with a fake story; she knows for sure I am disrespecting her by accusing her of lying.

This is the deployment of the sucker schema in weapon form, meant to poison the trust between two people. It launches a relationship paradigm shift: suspicion, not cooperation; verify instead of trust. Bill Withers complained (musically) about this very dynamic in "Use Me." Break up with your girlfriend; your love affair is a snow job. Your grieving student is a cheater. Take heed.

The Prisoner's Dilemma

My students and I have a power dynamic that comes with our institutional roles, but in other contexts people will infer status stakes and power plays, as we saw in the bare-bones social world of the Public Goods Game. In 1965 the mathematician John Nash decided to try out a classic sucker's theorem on real people. Nash is known for the Nash equilibrium, the foundational concept in game theory that says the game is "solved" when no player can do better by unilaterally changing his own move. At the time, Nash was working at the RAND Corporation, and the real people available to Nash and his team were the secretarial pool. The economists brought groups of secretaries in and introduced them to one of the most famous economics games of all time: the Prisoner's Dilemma.

At the level of economic theory, the Prisoner's Dilemma is just a contingent transaction with a payoff matrix to account for the choices of two independent choosers, sort of like a formalized game of chicken. The incentives are structured to pit the urge to cooperate against the fear of being exploited, because each player can always do better than the cooperative outcome by choosing to betray the other.

To make the theory more vivid, and intuitive, the game was described in terms of two prisoners. The story imagines two co-conspirators who have in fact committed the crime they are suspected of. They are held for questioning and interrogated separately. The DA tells each suspect that the standard punishment for the crime is ten years, but the state lacks the evidence to convict. The DA offers each a deal: Turn state's evidence (snitch) on your accomplice, and you'll serve no time at all while your accomplice serves the full sentence. If both suspects stay silent, they will only be able to convict on the lesser charge, a two-year sentence for each. If both suspects betray the other, though—that is, they both try to take the deal—they each get a reduced sentence of five years.

So the RAND economists brought the secretaries into separate rooms, matched them up randomly via ID numbers, and asked them to each indicate what they would do (though obviously with monetary payments rather than imprisonment as the actual outcomes): Cooperate or defect?

The economists thought it was clear what each person would and should do. Any given player would be better off snitching no matter what the other player chose. If my partner is going to betray me, I had better betray back and take five years; if I am quiet while he snitches, I would get ten full years, which is clearly worse. And if my partner is going to be quiet, I get two years if I do the same—but zero if I snitch. Again, I'm better off snitching. Which is to say that the Nash equilibrium is: both snitch, both serve five years.

The secretaries defied the prediction repeatedly. They cooperated (stayed silent) again and again, so much so that one of the researchers accused them of refusing to play by the rules. The secretaries saw the hierarchy, I think, in the same way as the decision researchers John M. Orbell and Robyn M. Dawes:

> The power of the [Prisoner's Dilemma] metaphor comes from the manner in which it captures the everyday understandings (1) that cooperative relationships can be productive to all parties involved—mutual cooperation is more productive in the aggregate than any other outcome; but (2) that exploiting another's cooperative behavior can be *more* rewarding to an individual than mutual cooperation; and (3) cooperative relationships are *risky*—there is always a chance of being "suckered" by a partner who "free rides" on one's own contribution.

For most people, there is real appeal in the best overall outcome but a deep fear that those attracted by the exploitative reward will make a sucker of the risk-takers. I get my most-preferred outcome if I choose to cooperate and my partner cooperates, too. I get my least-preferred outcome if I choose to cooperate and my partner defects. So who takes that risk? Who risks being suckered in the hope of group benefit?

Forty years after the secretarial revolt, psychologists ran the Prisoner's Dilemma again. As in other games, they handed the subjects a bland payoff matrix, a grid that would make it clear that the players could benefit from cooperation but benefit even more (maybe) by pulling a fast one. In this version, though, the subjects were assigned to read that the game was called either the Community Game, or the Wall Street Game. The Community Game elicited typically high levels of cooperation, up around 70 percent. But in the Wall Street Game, cooperation was half that. Same game, new packaging!

Calling a game the "Wall Street Game" is not specifically accusing the players of cynical motives, but it has the same effect. It acts as a prime, as in "priming the pump," where one idea automatically connects to related concepts. The gesture to greed or capitalism is an implicit warning that a sucker's game is in the offing, and the sucker's game follows different rules. The subjects who played the Community Game—and likely the RAND secretarial pool, too—would likely have agreed that cooperative, unselfish behavior was the moral choice, the "right" choice.

But one of the reasons that the sucker weapon is so easy to deploy is that it does not ask people to make the "wrong" choice. It tells them that they have misunderstood the rules altogether—the norm is not compassion, it is smarts. The rules of smarts are different, which two psychologists from Princeton, Dale T. Miller and Rebecca K. Ratner, laid out carefully in their work on the "norm of self-interest." They observed that there is a subtle social norm that says that society respects and admires people who are savvy and even selfish.

The idea of a norm of self-interest was a bit controversial: Why, critics asked, would you need to talk about a "norm" of self-interest? It's like saying that people drink water because there is a norm of hydration. People drink water because they are naturally thirsty; it's not a social phenomenon that requires a norm to perpetuate. Similarly, we might think that people act in their own material interests—try to win or try to profit—because it's what people want individually; indeed, it's tautological. Surely the reason we have to articulate the Golden Rule, and call it that, is because we are trying to dissuade people from their natural inclination to selfishness.

But if you think about the Ratner and Miller hypothesis in the context of the sucker, it becomes clear. The mark—credulous, agreeable, tenderhearted—is a fool, not a paragon of virtue. A society may have a strong norm against stinginess but nonetheless insist on respect for the savvy. People fear "dismay, suspicion, or derogation" if they make themselves vulnerable to economic exploitation.

The norm of self-interest affects behavior in surprising ways. Americans, who are quite charitably generous at baseline, appear to be more comfortable describing charitable giving in terms of supposedly selfish motives rather than just announcing their own altruism. They explain their contributions as personal investments, like donating money to medical research because your family member suffers from the particular medical condition. Ratner and Miller were especially taken with the research of the Princeton sociologist Robert Wuthnow, who had seen these altruistic disclaimers when he wrote the book *Acts of Compassion*. He reported that many charitable volunteers decline to have their good deeds characterized as compassionate, and explain their generous voluntarism with excuses like "It gave me something to do" or "It got me out of the house." People are nervous about being seen by others as a "bleeding heart" or a "do-gooder"; Ratner and Miller showed that decisions made by groups—that is, decisions in a social context—tend to be more selfish than decisions made by individuals. Those who take the norm of self-interest seriously act more selfish in public than they would prefer if they were unconstrained by social pressure.

When the Prisoner's Dilemma was renamed the Wall Street Game, subjects picked up the norm signal: the rule here is smarts; don't get fooled.

In *Dirty Rotten Scoundrels*, a movie we rented a couple times a year in my childhood, Steve Martin's con man is in the dining car on a European train and ostentatiously orders himself only a water, luring a kindly woman to take pity. He is saving money, he tells her piously, for his sick grandmother. She was the woman who taught him that "it is better to be truthful and good than to not." In my family, this line slayed every time. The banality! The tautology! The irony!!! It occurs to me now that we didn't get the joke. The sucker rhetoric asks: What if it is *not* better to be truthful and good? What if that just makes you a chump?

Every day, we make hundreds of decisions that implicate social values: Do I share? Do I give? Do I make space? The selective triggering and defusing of social norms has serious consequences for behavior. Invoking the possibility that you might be judged a fool means, perhaps, you don't let the guy merge in front of you on the highway, or you don't leave the bigger tip, or you don't cover for your coworker who is leaving early again. These smaller crossroads are emblematic of what happens at social, cultural, and political levels. How does fear of being played for a fool affect our appetite for humanitarian aid to foreign countries? Our understanding of compelling asylum claims? Redistributive policies, the franchise, education—these arenas have all been subject to the claim that people are trying to game the system with welfare fraud, or voter fraud, or phony addresses. This is sugrophobia in place of civic virtue, where the urge to protect the vulnerable is hampered by the fear that one undeserving beneficiary makes a patsy of everyone else.

At its most potent, the fool's game is a kind of biological weapon, invisibly contaminating even the most intimate familial circles of empathy and trust.

When my son was ten, he started fifth grade in a new school. He found the homework reasonable, and he liked his teachers. He ran for class vice president, explaining cheerfully to us that it was actually just his own homeroom, not, like, the whole grade. He loved his music instructor, a wry Beatles fanatic, so he joined the middle school chorus.

In October he went on a camping trip in western Maryland with my husband. A few weeks later he came home from school with a stomach flu. Days passed and he couldn't shake it. By the end of November we had been to the emergency room three times and he had been admitted for inpatient tests. He missed school or called home from the nurse's office day after day. Or he stayed at school and came home on the bus, stumbling in sweaty and pale. He was a husk. It was the worst season of my life.

About a dozen doctor's visits in, we had no diagnosis, but we did start to get what I would charitably describe as parenting advice. Perhaps we should not be so solicitous of his complaints of pain? We would encourage "sick behavior." Had we considered a less stressful school? Many children have "tummy aches" when they are anxious about their grades. (I should have walked out of the office the minute an adult used the word "tummy" to explain something to me, another adult.) One pediatrician would prod my son's abdomen, and if he grimaced inconsistently, the doctor would raise his brows and widen his eyes at us over his patient's head.

After one such visit, the last of this kind, a picture snapped into focus for me. They thought *I* was the sucker! They thought my fifth grader, consciously or not, was *scamming* me!

Briefly, the doctors' condescension did exactly what a sucker threat is supposed to do. I felt sheepish, diminished. I think we were actually advised not to ask how he was feeling so we wouldn't *remind him to feel bad*. I looked at my miserable child and felt a fillip of doubt.

I really have only one job in this world—to get my kids through to adulthood—and for a minute there I almost disengaged. Ultimately, it was helpful to suddenly apprehend the implicit sucker accusation for what it was so that I could inspect it and reject it. And we did move on; we saw another doctor who asked more questions about the camping trip, the ticks, the sore joints. Tests were run, bloodwork puzzled over. We funneled him handfuls of antibiotics and he recovered, slowly, like a child in a Victorian novel.

The whole episode was terrifying, obviously. My son is now the tallest person in our house, a funny and independent teenager who plays Wordle and blends voluminous protein smoothies right as I am trying to go to sleep at night. He's okay! And I am still haunted by the near miss. What if I had taken the belittling to heart and stopped looking for answers? What if I had let myself be poisoned against the person who needed me most?

Surveillance

It is not a coincidence that the suspected schemers in so many stories are students, children, immigrants, girlfriends, upstarts, or prisoners. Although these are not the scammers who pose the biggest material threat, their scams, even in the hypothetical, have enormous stakes for the social status and psychological self-regard of the would-be mark. Economic games might suggest that fear of the sucker is a universal phenomenon, even if you strip away cultural markers and status cues. This is true, but it can create a misleading picture, because the fear of exploitation is not evenly distributed across potential scammers.

At a primitive emotional level, the retort to a threat of a scam is: How *dare* you? But you can also think of the sentence with a different emphasis: How dare *you*? If you are scammed by a subordinate—my own kid! my student!—it makes you a loser. That visceral recoiling at the language of dupes—chumps, stooges, boobs, suckers—is a frantic desire to get some distance between the self and the bottom of the barrel. The force of the rhetorical weaponry is that it alienates and insults everyone involved. With one gesture, the doctor can tell me: your kid is a liar and you are a fool. A little bit of throwaway snark lets my colleague imply: students don't deserve respect and neither, apparently, do you.

It is also a remarkably sophisticated weapon. Take the student example. If a student puts one over on me, I feel stupid, and I know I look weak. The threat from the student feels very acute to me, at least in terms of psychological triggers, because it is a big status threat: I don't want to be seen as weaker than the people I'm supposed to be leading. At a primitive level, it feels like a bigger deal than being conned by, for example, the provost's office or the bank, because being undermined by my own students is really bad for a professor. When I can play all this out in my mind prospectively, it changes how I behave toward those same students in anticipation of the harm they could do to me. It makes me vigilant and suspicious toward people who would presumably find my stance confusing, since by all other accounts I am the less vulnerable party in our relationship.

This interplay of power relationships and causative actions can be written in quasi-mathematical form as well, almost like a logical proof. It feels a little formalistic, but it helps string a couple of key things together, so I'm going to try it.

Let's say we have two figures, the grifter and the mark:

1. The grifter's successful con moves the parties' power relationship, elevating the status of the grifter and debasing the credulous mark.
2. The lower the initial standing of the grifter relative to the mark, the more threatening the con, because the mark falls further when conned by a low-status grifter.
3. If more resources are, naturally, invested in preventing the more threatening con, more resources will be deployed to police potential threats from lower- rather than higher-status grifters.

Together, these propositions yield a cultural hypothesis: the threat of feeling like a sucker motivates people to defend vigilantly against incursions from the lower rungs of economic and political power. This is true at the individual level and it is true at the level of systems. In the United States, the people suspected, structurally, of scamming the system are not the powerful. They are the earned-income tax credit filers, making less than $20,000 per year but more likely than millionaires to be audited. They are

Amazon workers subject to biometric monitoring during their delivery shifts. They are the residents of Ferguson, Missouri, where the Department of Justice found an intensity of police surveillance that subjected Black residents to a quasi–police state.

Indeed, the American racial hierarchy is an especially salient exemplar of this dynamic, and one that I will take up in more depth in the chapters that follow. But it does not take a full-length treatment to point out that the American institutional commitment to white supremacy yields a relentless suspicion of Black everyday activity.

At the level of the individual, echoing a bit of Dolly Parton, the sociologist Tressie McMillan Cottom described the dynamic of who gets flexibility at work: "Ask a middle-class, white-collar Black woman how easy it is for her to call in to tell her colleagues she will be working from home today and see how she stares at you. I was the most published and, arguably, successful academic in my previous department. Yet as I walked down the hall one day, a white male student-worker felt empowered to tell me I should work with my office door open to prove that I work."

This description of individualized surveillance—if you are Black, your boss wants to see what you're doing and the intern thinks he deserves to see what you're doing, even if you are the most famous contemporary sociologist in America—plays out as a micro version of the excess surveillance that characterizes the Black experience. In academic and cultural commentary, the experience is often referred to as Living While Black—Shopping While Black or Driving While Black—a persistent reminder that racial hierarchy is enforced with blanket suspicion.

And, of course, being openly suspected of running a scam is itself alienating. When my grieving student learns that I need documentation to verify her bereavement absence, she is learning something about her standing and she feels demeaned, and rightly so. My casual suspicion tells her that she is not part of the group that gets deference or the benefit of the doubt.

The heightened fear of sucker threats from below has the perverse consequence of making those who wield less structural power more relentlessly subject to suspicion and surveillance. Which leads to the final link in the logical chain of weaponization, that surveillance itself is a means of denying claims to equal status:

4. If low-status grifts are the ones that get policed, then being surveilled, or suspected, is itself a social insult. It conveys that the social capital of the suspected grifter is low enough that the power play is notable and unwelcome.

If I, in some administrative capacity, were to ask a *colleague* for documentation of her stated reason for missing some event—say, a faculty meeting—it would be a relationship-ending conversation, so deep would be the offense. The thought of this is so outrageous that it makes me uncomfortable just writing it. (Once, briefly, there was a sign in our faculty lounge announcing that the refrigerator was under video surveillance because people were taking too many sodas. Faculty, who knew perfectly well that there was no video and who in any case were not the targets of the sign, were nonetheless beside themselves at the insinuation. The sign was removed.)

The broader structural point has been made carefully and thoroughly by sociologists of race. Many scholars have documented the Black experience of over-surveillance as a means of derogation. Surveillance in this context is specifically a weapon for social control, keeping close watch over people who claim to be doing something pro-social—grocery shopping, working, driving, letting themselves into their own house—but are categorically under suspicion that they are actually trying to cheat the system.

Even those otherwise in favor of over-surveillance for others understand the status implications immediately when the camera is turned back on themselves. After the murder of Michael Brown in Ferguson, Missouri, and subsequent protests, the Department of Justice conducted an investigation and issued a lengthy indictment of the city's policing. Among other horrifying examples, the

Ferguson report described an officer's domineering rage after he stopped a family on a trivial pretext:

> The mother then began recording the officer on her cell phone. The officer became irate, declaring, "You don't videotape me!" As the officer drove away with the father in custody for "parental neglect," the mother drove after them, continuing to record. The officer then pulled over and arrested her for traffic violations. . . . When the father asked the officer to show mercy, the officer responded, "No more mercy since she wanted to videotape."

There was no neglect and no traffic violations. There was, however, a frenzied, reactive cruelty in the scene, with echoes of Trump's rageful insistence on the lie of birtherism. Is the white working class—everywhere, and in suburban Missouri—going to let Black people put one over on them? The job of surveillance is to enforce subordination; the potential grifters (jaywalkers, parking violators) are prevented from leveraging even minimal power.

Surveillance itself weaponizes the sucker dynamic, reminding everyone of who is under suspicion and who is above it. Sending security guards to follow around a shopper; pulling over a driver—a family—on the thinnest pretext; surveilling the office supplies closet—these quasi-accusations are power plays. There are some people or institutions from whom it is acceptable to take a bad deal, and some from whom it is not. There are some scams that are so deeply embedded in the economic and social system that they are not reliably policed as exploitation at all. Those who are so policed know exactly what that vulnerability implies, about themselves and their social capital. Or, as the arresting officer, a literal enforcer of a police state, told the handcuffed parents: "Nobody videotapes *me*."

CHAPTER 3

Flight

Now that I am the parent of adolescents, I have been introduced to a variety of exciting new social categories. A recent favorite is the "try-hard," a punishingly literal pejorative for people who, well, try hard. For the uninitiated, a try-hard is a nerd, but even more earnest. As shorthand, it is an elegant repudiation of the parental party line; teenagers have always known that the biggest losers are the ones who tried to win in the first place.

At a fundamental level, the sucker tries hard. He invests, gets involved, sticks his neck out, goes all in. A sucker is a naïve cooperator, someone who plays by the rules while everyone else gets the prizes.

The instructions for try-hard redemption follow logically from the insult itself. Fools try hard; you don't want to be a fool; don't try hard. You can't be a sucker if you refuse to engage in the first place. For people afraid of being scammed, that looming threat can be skirted, sidestepped, neatly avoided. Often the avoidance is intentional and sensible. My email spam folder, for example, currently holds many offers for lucrative engagement with international finance, international romance, and charitable ventures. Although I am interested in profit and charity (less so extramarital romance), I am going to delete these messages. I could write back and suss out the details, but the odds that a Moldovan prince is holding money for me in a trust are discouraging. Even for pitches that are a closer fit for my interests—wind energy! Bitcoin!—the whiff of a Ponzi scheme is a helpful counterweight.

Other times, however, the urge to steer clear is more primal and less deliberate, an instinctual fight-or-flight response to fear. My kids love to repeat the story of a family road trip in western Maryland when I, from the driver's seat of our minivan, spotted a long black snake curled up on the side of the dirt road. I saw the snake start to uncoil maybe thirty feet ahead. I braked abruptly, and . . . rolled up all the car windows. (Then I gathered my wits and drove slowly by the unconcerned snake while my spouse and ungrateful children workshopped their impressions of me for three merciless hours back to Philadelphia.) The adrenalized reaction that we might have when we approach something scary in the wild—a mountain lion, a bear, a snake—can also take hold when we spot the looming grift. Although the idiom lists the "fight" first, in the real world it's the flight response that gets the most use, that inchoate instinct that says, *Get me out of here.*

In everyday life, people are not literally fleeing fundraisers or email scams. Instead of running away or crossing the street, flight from a con can be targeted disengagement, or refusal to play along, or just passive inaction. It's just a frisson of fear that triggers a quick step back. In a Public Goods Game, the wary will refuse to throw their money into the pot; on a dating app, the suspicious might swipe left on a profile that seems too good to be true.

But when this elemental response to sugrophobia avoids vulnerability altogether, it can sometimes redound to the detriment of the self and the social fabric. It's not just refusing to invest in Bitcoin; it's refusing to invest in social support or civic cooperation.

The fear of playing the sucker distorts our preferences for cooperating, at the personal and at the political level.

Don't Cooperate

From a social science perspective, it can be difficult to distinguish
between normal avoidance and sugrophobic flight.

One of the functions of an experimental social science like psychology is to exclude competing explanations for the same observed
phenomenon. Especially when the "observed phenomenon" is
someone *not* doing something, there are many reasonable accounts
of what's (not) happening. Imagine that a professor from another
school sends me a message inviting me to contribute a chapter to
the book he's editing on contract law, and I just let the email sit in
my inbox. Let's assume that the editor gets paid for the book but
not the contributors; let's also assume that I suspect I have been
invited because they needed a few women's names sprinkled in
rather than an interest in my work per se. Did I fail to respond to
the offer because I was too busy? Because I don't like writing book
chapters? Or because of a dim foreboding of suckerdom? It can be
hard to know—and, even more challenging, I'm not sure I could
accurately report my own motivations. Some of that processing is
going on below the level of conscious thought, so it is very difficult
to pinpoint.

A well-designed randomized experiment can whittle down the
possible explanations by manipulating the incentives to show, for
example, that even if I was interested, and even if I had the time, I
would still be wary.

With these goals of causal inference in mind, in 1983 the psychologist Norbert Kerr set out to show how subtle interpersonal
cues could derail a cooperative activity for the sugrophobic.
He took as his working model a common sucker's dilemma: the
group project. If you have ever been part of a challenging team
dynamic, the setup will feel familiar. You and your classmates

or colleagues are tasked with some sort of joint venture, like a presentation. You personally are working hard and want to do a good job. At some point you look around and realize you are doing most of the work while everyone else coasts. You really wanted the good grade or the professional acclaim—but nonetheless you ask yourself: *Should I just do this all myself and feel like a chump, or should I slack off on principle?* Refuse to work and you'll bomb a project you care about, but maybe it's worth it to thwart a free rider?

Of course, there is something slippery about that change in priorities, something petulant about refusing to contribute "on principle" to a project you actually care about, just to make someone else as unhappy as they deserve to be. To spotlight the sucker slack-off effect, Kerr's studies experimentally manipulated the costs and rewards of working versus shirking.

It would be unethical to randomly assign students to be more or less lazy on graded group projects just to measure their partners' responses, of course, so he devised an alternative task suitable for a controlled experiment.

The procedure went like this. Kerr invited students into his lab and seated each one in a curtained booth. He told them that they could earn money by performing well on a motor task and then pulled back the curtain to reveal a table equipped with a "microcomputer" (keep in mind this was in 1983), assorted audio equipment, and a box labeled FLOWMETER. The box had a Rube Goldberg quality to it. Two color-coded hoses were attached to the sides, and at the end of each hose was a rubber bulb (imagine the kind you use to clear a baby's sinuses). Each student was assigned a hose and then introduced to her partner. Crucially, the partner was actually a "confederate"—another student who acted

---

* The description of this study in the original paper evokes nothing so much as the opening scene of the original *Ghostbusters* movie. One imagines Bill Murray waggling his eyebrows while coeds are earnestly squeezing air into a "flowmeter." And speaking of dupes, we would be forgiven for wondering exactly what kind of participants looked at this setup and thought: *Yes, seems legit.*

like a research subject but was secretly in on the study and following a script.

To start off, the experimenter told students to grasp their assigned rubber bulbs and then squeeze them to pump as much air as they could in thirty seconds. The more times they squeezed and released, the more air was pumped, like pumping up a bike tire. On the box, situated above each hose, were two small amber lights, one for each subject; a subject's light would flash every time she "succeeded" at the task. By the rules of the game, subjects who pumped 350 milliliters of air had succeeded, a target that would determine their eventual bonus payments. Each subject could see their own light and their partner's light, so they had full shared knowledge of who was doing what.

They were told to do four practice trials before the paid rounds. The real subjects—the people whose behavior was being studied and who were responding spontaneously—practiced as instructed. The confederates, however, worked off their script. Some confederates performed the task reliably, reaching the goal in three out of four rounds. Others had been instructed to perform the task poorly and met the target in only one round out of four.

Overall, this meant that half of the naïve subjects went into the paid rounds thinking they had a partner who was capable of flowmeter success, and half went in thinking their partner was a naturally weak squeezer. With this background understanding, the subjects were then presented with the rules of the paid rounds.

Both partners were instructed to simultaneously attempt to reach the 350-milliliter goal in thirty seconds as before, but now in a set of ten rounds. Most importantly, for every round in which either one or both partners hit the target, both partners would be paid a bonus. This meant that if I were a participant and I reached the goal, I would get a bonus; if I failed but my partner succeeded, I would get a bonus, and if both of us succeeded, I would get a bonus. The only way to get no bonus was for both of us to fall short.

Now, by design, it was easy to pump enough air to succeed.

When students were left alone and asked to try their best, success rates were over 90 percent. So whether they had a helpful partner or an unhelpful partner, most people could get the bonus payment almost every time, because only one person needed to hit the goal for the bonus to be triggered. But Kerr hypothesized that subjects would change their effort depending on their partner's behavior. To set the right conditions to test the hypothesis, he told all the confederates that once they got to the paid rounds, they should stop being helpful; in other words, they should purposely fail the task.

In theory, given that everyone could individually squeeze enough to get the bonus, it shouldn't have mattered how or whether the subjects were partnered. But Kerr's prediction was that subjects would respond to their teammate's behavior in sugrophobic ways, downgrading their own effort when they spotted a scam.

Some subjects had a partner who was weak in the trial rounds and still weak in the paid rounds. But others had a partner who was fine at the task in the practice rounds and then suddenly performing poorly in the paid rounds. If you were the subject in that case, what would you think had happened to your hapless colleague?

Kerr's intuition was right. People paired with a *weak* partner earned the bonus about as often as if they had been working alone. But when paired with a *shirking* partner, subjects didn't even try for the bonus, apparently avoiding the possibility that a "win" could turn them into a sucker. When the lazy partner made the ridiculous squeeze task into a sucker's game, the cash bonus wasn't worth the psychological price, and they withdrew. *Roll up the windows, stop squeezing the bulb: I want out.*

Don't Invest

In some ways, Kerr's subjects were lucky: they could dip their toes in the cooperative waters gradually, trying to make the partnership work, but backing out before they were stuck playing the fool. They didn't have to *trust* that their partner would try hard; they could respond to laziness by withdrawing their own effort, tit for tat.

The sucker stakes are even higher when the potential chump has to make the first move. Most investments, from the financial to the professional to the intimate, are not simultaneous cooperation. One party takes a leap of faith and the other responds, even in just regular day-to-day ways. You leave your kid at day care the first day and hope they come back clean(ish) and fed and happy. You loan some money to a friend and wait to get repaid down the road. You say "I love you" first and hold your breath.

Or . . . you don't. The prospect of trusting and being betrayed is aversive enough that many people will disengage before they even start. Trust is emotional cooperation, with high psychological stakes: it's not just my money on the line, it's my sense of self.

In the summer of 2020, I wrote a short article for a national magazine. After the publication, I received a few emails from strangers who claimed to be literary agents interested in speaking with me. I deleted them, just like I delete any number of email offers of dubious "opportunities." Most of the offers I decline-by-trash-folder are ridiculous, fake, and unrelated to my personal goals. By contrast, I really did harbor secret writerly ambitions. Having an agent sounded very, very cool to me. But when I saw those emails, I thought: *If I reply to this and it's a scam, I am going to feel like such an idiot. It's not worth the trouble.* (A few days later I went into my trash folder and clicked Move to Inbox on one of the messages. It was not a scam, and that's how I got to write a book about suckers. I see the irony.)

What was I so afraid of? Admitting to an internet stranger that I had visions of artistic grandeur made me vulnerable in weird psychological ways. If it worked out, something amazing might happen. If my trust were betrayed, though, I would look and feel like a vain fool who couldn't resist a flattering come-on. (Sure, I'd love to dance, thanks! Oh, I see, never mind.)

You can only be betrayed if you trust in the first place, and betrayal feels awful. Like regret, it carries an emotional lesson that gets overlearned; people fear their own future dismay so intently that they distrust in anticipation. In the language of psychologists, the reluctance to risk misplaced trust is evidence of "betrayal aversion," and it is acute enough for some people that they even attach it to *objects* that they think might turn on them. In fact, one line of research has linked betrayal aversion to vaccine hesitancy, where the threat of serious side effects is sometimes perceived as a medical betrayal. The vaccine that promises to help but instead does harm can look more menacing than the illness that promises harm and delivers.

As a matter of social science, research on distrust is challenging, because it's the study of things that don't happen. The refusal to trust is a dog that doesn't bark; it doesn't leave a trail because there's no affirmative record left when someone sidesteps an interaction or passes up an opportunity. It is hard to observe the George Fournier types, the suspicious Mainers who don't trust their money to a bank, because the whole point is they are hiding it at home, intentionally unobservable. In the real world, it's impossible to measure all the trust that doesn't happen.

Perhaps for this reason the behavioral study of trust has been heavily dependent on economics games. We have already seen cooperative games—the Public Goods Game, the Prisoner's Dilemma—that have a leap-of-faith flavor for players who want to cooperate. But in those games, the players make their choices simultaneously and blindly. Betrayal, by contrast, comes from an implicit contingent promise. There are two stages: If *you trust me,* then *I'll be trustworthy.*

What the Public Goods Game and Prisoner's Dilemma experiments lacked was a contingent structure, a way to let one player venture the first "I love you" and see how the other player would respond. The protocol that eventually emerged to study the "social aspect of risk," as they quaintly put it, was developed in 1994 at the University of Iowa by Joyce Berg, John Dickhaut, and Kevin McCabe. Originally dubbed the "Investor Game," it soon became universally known as the Trust Game, and you will see why. It is one of the most foundational experiments in behavioral economics.

In the standard Trust Game, there are two players, whom we usually call the Investor and the Trustee. The Investor gets the starting money, an endowment of $10.*

The game has two moves. The Investor moves first, when she is invited to pass some of her money to the Trustee. She can pass any amount between $0 and $10. Whatever money the Investor passes along is automatically tripled, so when the Investor shares, it is really lucrative for the Trustee: if the Investor shares all $10, the Trustee gets $30.

If the Investor takes the leap and shares some, she then has to wait to see what the Trustee will do. In the second and last move of the game, the Trustee is invited, but not required, to send something back to the Investor. Having seen what the Investor chose to share, the Trustee gets to decide how much of a return the Investor gets on the investment. Once the Trustee chooses how much to pass back, the money is split according to his instructions and the game ends. (As with all these games, the players never know each other by name or see each other in person.)

The Investor's incentives are surprisingly subtle. The Investor knows the only way to make extra money is to share it. Sending it is also the only way to be generous: if the Investor doesn't pass any money, the Trustee goes home with nothing. But, of course, whether and how much to send is fraught, because the Trustee will have to react and respond.

Just as in the Public Goods Game and the Prisoner's Dilemma, some people cooperate and some people do not. In the original version of the game, out of thirty-two pairs, five Investors sent all $10 to their assigned Trustees; the majority sent between $5 and $10; and two sent nothing. This basic pattern has been replicated many

---

* As in other games, the amount of the endowment changes from experiment to experiment but is often described in units of ten. For simplicity's sake, I again shorthand it here as $10.

times since then, and it reflects a range of ways that Investors—
just like any of us shopping or negotiating or taking a gamble—
are working through the question: *Is this guy going to screw me?*

One of the challenges for interpreting behavior in the Trust
Game is that it can be hard to disentangle different kinds of fears.
One argument against the sucker interpretation of Investor reluctance was that some people are just more risk-averse. On that view,
it's not that they are worried about being betrayed or duped specifically; it's just that they prefer not to engage in risky behaviors
generally. In 2003 a team of researchers led by Iris Bohnet and
Richard Zeckhauser at Harvard wanted to show that it was not just
any risk but the particular risk of betrayal that the Investors were
afraid of.

With some of the details simplified, here is the idea: Imagine
that you walk into the lab, and as soon as you do so, you get your
$10 endowment, just as you would if you were chosen as the Inves-
tor in a Trust Game. You learn, however, that the Trustee you're
paired with does not get to make her own choice in this game.
When it comes time for her to pass the money back to you, she's
going to have to do what a computer tells her. The computer just
rolls the digital dice and, depending on what comes up, does one of
two things: doubles your investment (high return) or sends back a
single dollar (low return).

Like all Investors in a Trust Game, you can just take the $10 and
go home. Investing is always a gamble and maybe you don't want
to gamble. But let's assume that the research team tells you this:
the computer's algorithm chooses high return about 80 percent of
the time, and low return about 20 percent of the time.

So the Investor, holding that $10, is thinking that if they choose
to gamble, they'll have a four-in-five shot at going home with $20
and a one-in-five chance of taking home $1. Would you take the
risk? I think I would. The average payout of that gamble (the "expected value" in economics terms) would be over $16, compared
to the $10 sure thing.

What the researchers wanted to know from the Investors was:

What are the *worst odds* we could offer that would be good enough to get you to take the gamble? This is called the MAP, the "minimum acceptable probability" for that high payoff.

The Investors were instructed as follows: You tell us your MAP, and we will compare it with our computer's algorithm. If the computer is giving out high returns more often than your minimum, we'll roll the dice for you. If it's lower odds, no gamble.

Imagine that the computer chose high return about 60 percent of the time. I personally would probably set my MAP at 50 percent. Since I was willing to take a worse bet than the one on offer, I would be given the gamble. If my MAP was higher than that—say, if I needed a 90 percent chance of winning in order to be willing to risk my $10—then the game would stop and I would take my $10 and go on my merry way.

There was, of course, an experimental twist in Bohnet and Zeckhauser's model. As I have described, some of the Investors were matched with a passive Trustee who could only do what the algorithm instructed. But some of the Investors were matched with an active Trustee, someone who could make a choice about whether to opt for high or low return.

Now the sucker stakes started to creep in—a real person who could betray a trusting Investor—and the researchers wanted to measure their effect.

As a group, the active Trustees were polled before the game started. They were asked: If your partner gambles, will you pass back $20 or pass back $1? All of the Trustees responded to the question, and they were all bound by the game rules to do what they said they would do. This means that the researchers could calculate the probability that an Investor paired with an active Trustee would get a high return: the probability was just the percentage of polled Trustees who reported they would pass back $20.

As before, they asked all of the Investors: What is your MAP for taking the gamble? That is, what is the minimum probability of high return that would make you prefer to take that risk rather than go home with your sure-thing $10?

The Investors did not know the odds that the computer or the poll had shown. They were just deciding for themselves what odds of a high return they would need to make the gamble worth the possibility of loss. The monetary harm from losing would be the same for subjects in either the active or the passive condition, whether they got a bad digital roll or a selfish Trustee. Bohnet and Zeckhauser predicted that even though the monetary harm was consistent across conditions, the psychological harm was different.

If the *computer* was randomly setting the odds and put the odds of winning (high return) at 40 percent, most people chose to take that bet. But if a *human* had a 40 percent chance of being trustworthy, that was not promising enough to inspire the Investor's leap. Being let down by a human seemed worse, and subjects tried to avoid it.

The decision-maker saw herself as a potential sucker and discounted the expected utility of the gamble by the risk of being exploited. She passed up an investment—one with a risk profile we know she otherwise approved—to avoid feeling duped.

The Wharton professor Adam Grant has made a broader argument about the costs of this phenomenon. In his book *Give and Take*, Grant surveyed a remarkable range of studies and examples of people whose financial success was tied to their outsized generosity. Grant named three styles of collaborative workers: the takers, the givers, and the matchers. His first observation is intuitive if you have ever worked on a team: takers face challenges succeeding at work. Workplaces are social environments and most people are quickly frustrated by, and skeptical of, especially selfish colleagues. More surprising, though, were his findings about matchers and givers. Givers are those who are generous even when they expect nothing in return. Matchers are those who operate on a theory of reciprocity. It seems like being a matcher would be a sensible middle-ground choice, not too easily taken advantage of and not too cynical to play well with others. But Grant observed that people who are afraid their favors will not be reciprocated—that is, those afraid of being suckered—just do fewer favors overall. They help less, and in the end, that means many fewer relationships and opportunities. "When matchers give with the expectation of receiving, they direct their giving toward people who they think can help them," he argued. ". . . At its core, the giver approach extends a broader reach, and in doing so enlarges the range of potential payoffs, even though those payoffs are not the motivating engine."

The studies about trust are often about risks that are easy to measure—specifically money—but in real life the risky investments are everywhere, from professional ambitions to financial planning to love. Friendships and intimate relationships are especially prone to these kinds of fears, because they have the potential to be rewarding but also come with real threats of the "social aspects of risk," like the fear that a friend will divulge your secrets or fail to reciprocate your kindness, or that a romantic relationship that promises love will not deliver.

Brené Brown, the University of Houston professor and bestselling author of a series of books on leadership and personal growth, writes about that inclination to disengage to avoid vulnerability. In her view, the disengagement has serious personal costs, preventing people from growing, contributing, and ultimately leading. In *Daring Greatly* she writes: "We disengage to protect ourselves from vulnerability, shame, and feeling lost and without purpose. We also disengage when we feel like the people who are leading us . . . aren't living up to their end of the social contract." There are risks we are willing to take only as long as a failure isn't also a betrayal—but of course that means that the fear gets a lot of power to ward off otherwise valuable opportunities. It is a costly fear; there is a reason that disdain for the try-hard wanes in adulthood. You can't lose if you don't try—but you can't win, either.

Don't Help

If you have ever seen the movie *The Silence of the Lambs,* or if you have ever been a child or a woman cautioned about men in white vans, you probably recognize a set of warnings about people claiming to be in need. *The Silence of the Lambs* in particular plays heavily on this trope to introduce its victim: she just keeps helping until she has helped herself into the recesses of a killer's windowless moving truck. In a dark parking lot, the unsuspecting young woman sees a man with one arm in a cast trying and failing to lift an armchair into a van. She offers to help him, though she herself has an armload of groceries, and they get the chair in. He asks her to get it all the way to the back of the cargo area, and she agrees—again—to climb in and push. Then he hits her over the head and pulls off the fake cast and brings her to his homemade dungeon. He wasn't injured; he didn't need help; he appealed not to her greed or her vanity but to her compassion.

Certain kinds of trust risks are not about what you get out of them but what you hope to give. Just like investment choices, altruistic opportunities also pose a sucker threat. Outside of the extreme serial killer context, trust in a charitable enterprise can lead to real financial or personal loss. You give your credit card number to a nonprofit that turns out to be a scam, or you help someone who stumbles on the street and they pick your pocket in the shuffle.

More often, the threat is not of loss per se but of misplaced generosity. Imagine for a moment that my sister sends me money when I tell her I have an unexpected car repair. She finds out later that I used the money to buy a new TV; I'm postponing the car repair and taking the bus for now. Does she feel scammed? I think so, although it's kind of funny. Her goal was to make my life better and she did make my life better: Didn't she get what she wanted? The answer is no, but not because of the money she lost, but rather because it was procured with a lie. Now she feels insulted that I lied to her and foolish for paying for my television; the next time she won't be so willing.

Across domains, the fear of undeserving charity can have dramatic effects on helping behavior. Arlie Russell Hochschild sketched a striking profile of the confounding tension between altruism and suspicion in her ethnography of the American right. In the book *Strangers in Their Own Land,* she recounts the interviews she conducted in a Louisiana parish, a Tea Party stronghold and a community struggling with the environmental fallout of deregulated oil drilling. One of her subjects was Lee Sherman, an environmental activist and a member of the Tea Party. Hochschild recounted Sherman's approach to a sort of skeptical generosity:

> He knew liberal Democrats wanted him to care more about welfare recipients, but he didn't want their PC rules telling him who to feel sorry for. He had his own more local—and personal—way of showing sympathy for the poor. Every Christmas, through Beau-Care, a Beauregard Parish nonprofit community agency, he and his wife, "Miss Bobby," chose seven envelopes off a Christmas tree and provided a present for the child named on the enclosed card. ("The card tells you a child's shoe size. If the size is too big, we know the shoe is actually going to an adult and we don't give. But my wife spends money we don't have on kids.")

Sherman and his wife lived precariously on Social Security themselves and often relied on help from family to stay afloat. Their commitment to charitable giving was so serious that they donated even when it was a material personal sacrifice. But their giving was riddled with suspicion, their donations in-kind rather than in cash: even big-footed children were rejected for fear they might be adults, although one imagines adults desperate enough to submit their own information to the tree charity would also be grateful beneficiaries.

The same instincts that complicate generosity at the individual level can be disruptive at scale when helping looks like redistribution of wealth. "Lee's biggest beef was taxes," noted Hochschild. "They went to the wrong people—especially welfare beneficiaries who 'lazed around days and partied at night' and government workers in cushy jobs."

Redistributive policies are consistently hampered by the fear of parasitic others, the perverse suspicion that the deserving rich are being taken advantage of by the indolent poor. As a result, people across the political spectrum often support social policies that are both inefficient and dissonant with their own stated moral values. American social policy has long been haunted by the specter that hardworking taxpayers will be taken for suckers while grifters are permitted to free ride.

The tendency toward skeptical giving shows up across the board. Even the most benign and familiar community charitable enterprises—toy drives, food drives, Coats for Kids—evince an underlying suspicion. Food donations are especially uncontroversial; few object to feeding the hungry. Because people are willing to give food, the drives generate a lot of social good. But the quiet counterfactual is that if people were willing to give in cash the same amount they'd spend on cans of beans or boxes of pasta, they could waste less of the gift. When food is donated directly, it requires many individuals to separately dedicate time and money to pay retail prices for the food they guess will be most useful. Going to the store to spend $10 on canned vegetables, a bag of pinto beans, and a box of rice to donate is a worthy charitable endeavor, but bang-for-your-buck, giving a ten-dollar bill to the food pantry—or the recipient—creates more good. A soup kitchen can aggregate cash donations and pay wholesale prices for food easily combined for group meals. An individual who receives $10 in cash can purchase the food most likely to be happily consumed, whether they are buying for babies or adolescents or the lactose intolerant. It's a paradox: people who are going out of their way to help are reluctant to go less out of their way—same expenditure, no shopping!—to be more ultimately helpful.

In-kind food donations are exploitation-proof: a bag of rice can feed a hungry person but it is less likely to buy drugs or alcohol. It's not commonly sold on a secondary market. More controversially, the bag of rice doesn't feel like a free ride. It's not a good enough prize to threaten to turn the tables on the donor; no one with a free bag of dried kidney beans is coming out on top.

This was demonstrated more systematically in a study by the sociologists Colin Campbell and S. Michael Gaddis, who pointed out an odd disjuncture in American public opinion: Americans evince widespread support for "assistance to the poor" but largely disfavor "welfare." The researchers pointed out that people think of welfare as cash assistance and housing programs, but "assistance to the poor" tends to be food banks, soup kitchens, and shelters. The authors decided to investigate this experimentally. They gave subjects a short story:

> Michael and Jessica are in their early thirties. They rent a home in a working-class neighborhood and own one car. They support their eight-year-old daughter and thirteen-year-old son. Michael works as a general contractor, but business is slow. Jessica works part-time at a preschool. Their combined income is typically $1,700 per month. Their expenses are stable from month to month: $700 for rent; $100 for utilities; $200 for vehicle loan; $50 for car insurance; $100 for gas; $100 for medical expenses; $400 for groceries; $100 for miscellaneous expenses, for a total of $1,750 in expenses each month. In months when Michael and Jessica have less income or unexpected expenses, they postpone paying bills, borrow money from family members or local charities, or charge the expenses to a credit card.

Campbell and Gaddis asked all participants the following question: What amount of aid should Michael and Jessica receive? When they asked the question, though, they randomly varied whether it was posed in terms of cash welfare, vouchers (i.e., food stamps), or in-kind aid (i.e., housing assistance or childcare subsidies direct to provider). Every subject was reporting a dollar amount; the only difference was in what form the dollar would travel from state to recipient.

Subjects distinguished between cash and in-kind aid, and they were more willing to give vouchers than money. People who wanted to give over $250 per month in food stamps were only willing to authorize $210 a month in cash welfare. As one of the participants commented, "I don't agree with giving cash—too easy to abuse the privilege."

Across redistributive questions, the sucker pattern recurs. A constituency emerges and makes a plausible claim: *We are in need.* One natural reaction takes the claim seriously, weighs out what we owe each other as citizens in a community, calibrates a response. But then comes the warning: What if this is a wolf in sheep's clothing? Suddenly the ask—food, healthcare, immigration reform, housing—looks darker and has to defend itself against the charge that, actually, the social safety net is for fools.

In healthcare: Alabama Republican Congressman Mo Brooks echoed the sucker logic when he suggested that the American healthcare system was taking advantage of those who lived "good lives" by forcing them to pay the same high health insurance premiums as those who did not. "They're healthy, they've done the things to keep their bodies healthy. And right now, those are the people—who've done things the right way—that are seeing their costs skyrocketing." The complaint was clear: the problem with universal healthcare was that it is a cooperative exercise in which the benefits of the insurance scheme threaten to divert resources away from those who play by the rules (good diet, no smoking, regular exercise) and toward those who are lazy or cheating (junk food, nicotine, preventable chronic conditions). These same themes resonate across healthcare debates, especially when the recipient is not a white cis man: Why should *I* have to pay for your birth control when you could just abstain? Why should *I* have to pay for your obstetric care when having a child is a choice that you made?

Or, to put it as a wealthy acquaintance once did, shockingly, in the early days of the Obamacare debates: Why should *he* have to pay for "sex change operations" for incarcerated transgender felons? His example has stuck with me for a decade: He was obsessed with the notion that universal healthcare was a means of letting other people trick him. The fact that he chose one of the most vulnerable populations I can imagine to illustrate the threat *to him* was illustrative of the deeper Fox News fever dream logic. He invented a fraud and found it so terrifying in the abstract that it justified his decision to help no one. Better safe than sorry.

In immigration: Donald Trump accused Mexico of duping the United States by "not sending their best people," taking advantage of our largesse by off-loading their unwanted population. When foreign citizens arrive at American ports of entry seeking asylum, they, too, are asking for risky help. But Trump asked: Do they really need help or are they "only" seeking the economic benefits of American life? Do they want to assimilate or are they "rapists and murderers"? In immigration, this fear of the sucker is writ large.

And in housing: it is cheaper and more efficient to provide free, no-strings housing to remediate chronic homelessness, but it is really hard to convince people of a policy that seems to reward substance abuse and unemployment. In a 2006 article in the *New Yorker* called "Million-Dollar Murray"—so named for an alcoholic homeless man in Reno whose annual healthcare costs often veered into the seven-figure range—Malcolm Gladwell painted a particularly vivid portrait of efficient social policy in tension with sugrophobic intuitions. Policy analysts had begun to point out that it was far cheaper to just provide Murray, and those like him, with free apartments. Even at a cost of $10,000 a year, it would be an enormous savings to the city if it kept him out of the emergency rooms. Indeed, the difference was more than fivefold, even for less serious cases than Murray. Free private housing, the analysts argued, would forestall many of the most serious medical issues for the small population of persistently homeless addicts, which often involved complications of being very cold or being vulnerable to falls and traffic accidents. But, on some fearful logic, free housing for the highest-cost homeless people means "rewarding" those who do not follow the rules, at the expense of rule followers.

Nonetheless, many cities—including, most famously, Denver,

Colorado—tried to implement reforms, and they met with a persistent refrain of unfairness. "Thousands of people in the Denver area no doubt live day to day, work two or three jobs, and are eminently deserving of a helping hand—and no one offers them the key to a new apartment. Yet that's just what the guy screaming obscenities and swigging Dr. Tich* gets," wrote Gladwell. "... Being fair, in this case, means providing shelters and soup kitchens, and shelters and soup kitchens don't solve the problem of homelessness." Shelters and soup kitchens provide intermittent help but don't keep people like Murray off the street.

Gladwell identified the friction as a conflict between serious moral principles—we should favor aid for those who deserve aid—and an unusual, costly problem that could only be solved by violating those principles. But as long as money is fungible, this seems like a sucker-fueled moral mistake. If the city spends a million dollars on Murray, that's money it doesn't have for the "eminently deserving. " The only benefit to denying the cost-effective aid to Murray is that you don't have to live with the frustration of seeing him jump the line. John Hickenlooper, mayor of Denver at the time and innovator in housing policy, reported ruefully that he was still stopped at the grocery store by constituents disgusted he was allocating money to help "those bums."

This tension persists across domains. The fear of playing the sucker by helping people is an explanatory mechanism for when we prefer lower-efficacy welfare interventions when higher-efficacy interventions are available. Means testing, work requirements, verification, and even bureaucratic hurdles—there are a number of common tools ostensibly used to target delivery but empirically more likely to increase the overall costs of a program. But they make the programs feel more exploitation-proof, and that makes them more palatable.

Martin Gilens, the sociologist and author of the book *Why Americans Hate Welfare*, argued that we often misunderstand American

---

* Dr. Tich is a mouthwash.

attitudes to welfare. Most Americans actually favor more, not less, help for the needy. It may feel surprising given the popular discourse, but he points out that "year after year, surveys show that most Americans think the government is not doing enough (or not spending enough) for education, healthcare, child care, the elderly, the homeless, and the poor." So why is there so much opposition to welfare spending? "The most important single component is this widespread belief that most welfare recipients would rather sit home and collect benefits than work hard themselves." Welfare, which in the United States is racially coded as a Black benefit, is disfavored because it feels like a scam—even when welfare benefits vindicate otherwise widely popular social goals.

At a deeper level, there is something about the nature of helping itself that has a perversely shameful edge to it. It is a little embarrassing to be publicly helpful. The mere fact of other-regarding compassion raises the possibility of foolishness. When Americans donate to charity but claim it's for their own benefit, they are reflecting a covert ambivalence to generosity and cooperation. If you have ever been called a "bleeding heart," you know it's an insult, a reminder that there is something suspect about people who would allow their own organs to bleed just to help others. Indeed, the first time the phrase was deployed in its modern sense, as an insult to political liberals, was in 1938, in a newspaper column by Westbrook Pegler. "I question the humanitarianism of any professional or semi-pro bleeding heart who . . . would stall [an] entire legislation program in a fight to ham through a law intended, at the most optimistic figure, to save fourteen lives a year," he wrote. The law he found so ridiculous was an anti-lynching bill. Civil rights legislation is for suckers.

Don't Compromise

Many avoidance decisions are subtle, a quiet skirting of the perceived risk. You just don't invest or don't donate; they barely count as decisions. But there is one version of sucker avoidance that manifests more overtly and affects not so much whether we fail to engage but rather whether we actively disengage. Sometimes the decision to avoid the sucker's risk is an affirmative, almost aggressive, Taking My Bat and Ball and Going Home.

One of the insights of behavioral psychology is that the fear of playing the fool is a barrier to successful negotiation. It is the same dynamic as the "flowmeter" study: one person feels shortchanged and decides, *Nobody is going to win here; I'm out.* It can be explosive or it can be entirely mundane.

After I graduated from college, I briefly lived in New York, and my parents came into town from Maine to visit me. We had been walking the city for hours, going to museums, sightseeing, and everyone was hot and hungry. We ducked into a grocery store to get some snacks and regroup. My father got an apple, went to pay for it, and came back empty-handed. Did we know an apple here cost over *two dollars?* He could not, in good conscience, pay $2 for an apple! (It may sound as though my parents were hot off the turnip truck, but they were perfectly familiar with Manhattan. I did not expect this to be my father's line in the sand.) He almost certainly had more than $2 in his wallet, and he really did seem to want that apple. I don't remember what he ultimately found to eat, but I bet it cost more than $2 and I bet he didn't like it as much as he would have liked the apple.

The little interaction between my dad and the store—seeing the fruit, seeing the price, walking away—was a kind of mini-negotiation. They couldn't come to an agreement, so it was a very short negotiation, but the structure is still the same. Two parties wanted an outcome that required an exchange, and they each had terms on which they would trade. My dad wouldn't trade, though, because to him it looked like a racket.

Whether the transaction feels scammy or not depends a lot on the context. So, to be a little pedantic, if I go to the store and see a display of something I am interested in—a fancy seltzer, say—I might go check it out. Let's say I see that I can buy a six-pack of the artisanal New England seltzer from maple trees or something, and that's right up my alley. Then I see that it costs $12 for this six-pack. And I don't buy it. I like seltzer a *lot*, but not that much. That deal will fall apart because the store won't meet my price. Given the store's preference to sell the seltzer for no less than $12 and my preference to spend no more than $9 on it, the reasonable and efficient outcome is for me not to buy it. Fine! No sucker problems.

But now imagine a slightly different situation. I go into the store and I see that the maple seltzer is on sale.* Say I've had this seltzer before and actually like it a lot. Even better, I see that it's discounted today to $6. I bring it to the register and it scans as $8. I ask what's going on and the cashier says, "It's eight dollars at point of purchase and then a two-dollar mail-in rebate." A mail-in rebate? That I will never mail in? What are they going to do, send me a check for $2? What a racket.

Of course, even without the nonsense rebate, the seltzer is still, in theory, worth it to me. I would pay $8 or even $9. But now I'm incensed. I'll pay nothing—and get nothing, too.

Bargains fall apart all the time; there are many things I don't buy because I don't want them enough to justify the price. But failing to buy something that *is* worth it to me suggests a psychological obstacle, in this case the sense that if I buy this seltzer, I'm playing into their bait and switch. Rather than pay the sucker's price, I walk away.

This walk-away dynamic is relatively low-stakes when consumers are making nonnegotiable purchases and extremely low-stakes when we are talking about novelty beverages. But the stakes get higher when there are real negotiations, especially when it comes

---

* This is getting very esoteric and reminds me of a theory class I took where we had to read a paper by a British philosopher on the origins of the social contract. We kept getting tripped up over the examples, because the resources that the theoretical humans were always dividing up in their proto-society were "plover eggs and port wine."

In the interests of full disclosure I have definitely bought maple seltzer, and more than once.

to resolving conflict or settling legal disputes. In her doctoral dissertation at Cornell, the psychologist Victoria Husted Medvec made this vivid with a little academic fable:

> A professor of psychology and a professor of economics are sitting in the faculty club, engaged in a heated debate. They are discussing a recent negotiation between one of their colleagues and the provost regarding a cut in the level of funding for their colleague's position. The economics professor exclaims, "I cannot believe that George refused to accept the provost's offer and decided to leave to go to another university for $15,000 less than his current funding level. Why wouldn't he have preferred the provost's offer of a $7,000 decrease relative to his current funding?" The psychology professor remarked that she was not at all surprised by her colleague's rejection of the offer. "But he took an even greater loss," rebuts the economist. "That all depends on your perspective," states the psychologist with a grin.

The grinning psychologist in her story understood something that the economist did not: feeling betrayed by the employer has a real cost to the employee. You don't feel betrayed by a new job offering a low salary, because there is no underlying trust to betray. The new job can afford to offer less sometimes, because its low salary doesn't come with an extra cost of the perceived psychological breach. People will walk away to avoid feeling shortchanged.

Medvec used a set of hypothetical bargaining scenarios to test her hypothesis that facing an unexpected pay cut from your employer is a concession that people will pay money to avoid. She called the phenomenon "concession aversion" and wrote that "the fundamental source of concession aversion is people's perception that there is a breach in the current relationship and that the other party is attempting to exploit them."

People are highly averse to the feeling that the other side is trying to take advantage of them. In legal settings, this is a serious problem, because litigation itself is really expensive. The first legal paper I ever wrote, with my graduate advisor Jonathan Baron, was about the problem of divorce bargaining. When people get divorced, the legal bills can be devastating. Most people do not have a lot of money to spend on lawyers, and divorce itself usually makes both partners financially worse off, because it's just more expensive to have two households. Nonetheless, a lot of couples wind up in protracted negotiations and even litigation, draining their precious joint assets.

My not-very-deep insight in the paper was that a lot of people go into divorce bargaining already feeling burned. Especially in cases in which the divorce itself is caused by a breach of the marriage contract—adultery, for example—the idea that the parties are supposed to negotiate for a basically even split of the stuff can seem outrageous. If my husband has left me for someone else and then we are at the bargaining table and he blithely suggests that we sell the house and divide the money from the sale *equally*, I very well may think, *What kind of chump would I be to accept this sucker's payoff?*

When I asked research subjects to take the perspective of divorcing parties in different negotiating situations, that was exactly what I found: it was not just that they disagreed or had different goals from their hypothetical counterparty but that, when the proposal came from a cheater, they thought it was just unreasonable. In many cases, two people looking at the same proposal—"Let's sell the house and divide the profit equally"—would have totally different views on whether it was even in-bounds as a fair suggestion.

The sucker dynamic gets triggered when a deal feels unfair, but fairness is in the eyes of the beholder, and it often looks suspiciously similar to the beholder's best interests. Anyone who has ever negotiated a shared item of food with a sibling will recognize this phenomenon. If I got candy, I think it's fair to give you a small piece because this is mine and I'm just being nice. If you got candy, the right thing to do is to split it in half to share with me because there are two of us.

Linda Babcock and George Loewenstein, faculty at Carnegie Mellon, wrote about this phenomenon in an article called "Explaining Bargaining Impasse." They were trying to figure out what triggered people to take the bat and ball and go home from the negotiating table, especially since doing that would make everyone worse off. They had a sense that people get stuck, and they started looking for the sticking point.

They gave pairs of subjects some information about a car accident. Each subject had to play a role, either the plaintiff (the person who got hit) or the defendant (the person at fault). The job was to figure out how much money the defendant would have to pay the plaintiff, between $0 and $100,000. Pairs who resolved their negotiation quickly would get a bonus. Nonetheless, they found that many of the pairs took a long time and forfeited the bonus, and they found that almost 30 percent of the pairs failed to reach an agreement at all, losing a substantial portion of their payment for the study.

When they looked at the results of the negotiations, they focused on two data points. Each partner had been asked what outcome would be good for them personally, and then what would be the range of objectively fair outcomes. Obviously, people in a negotiation diverged on what outcome they thought would be good for them personally; plaintiffs wanted a high payment and defendants wanted a low payment. More strikingly, though, in a substantial subset of cases, the partners also diverged on their assessments of the objectively fair outcomes. In fact, for most of the pairs that were unable to reach an agreement, their ranges of "fair" outcomes did not overlap at all.

The self-serving bias led the subjects to the dispiriting conclusion that they were being scammed. As Babcock and Loewenstein put it: "If disputants believe that their notion of fairness is impartial . . . then they will interpret the other party's aggressive bargaining not as an attempt to get what they perceive of as fair, but as a cynical and exploitative attempt to gain an unfair strategic advantage."

When parties disagreed about what was fair—implicitly influenced by their own self-interest—it resulted in them perceiving the other person's bargaining as something more than just playing the game. It made the counterparty seem to be cheating, trying to take advantage. And, moreover, once the perception of unfairness was entrenched, it was fatal to the negotiation, because "negotiators are strongly averse to settling even slightly below the point they view as fair"—that is, once people get a whiff that they're being taken for a chump, they're out. It was literally preferable to lose big but stand proud than to lose small but play the fool.

The fear of playing the sucker puts our selves and our values into a fun house mirror, exaggerating some harms and minimizing some values. The fear changes our attitudes toward cooperation and giving and in turn affects how a society allocates resources for healthcare, housing, and welfare benefits. It asks members of a community to neglect each other, and when there are hard problems to be solved, or conflicts to be negotiated, it can make us quicker to throw in the towel.

This sugrophobic aversion to compromise has deep implications for the social order and also a deep layer of complexity for the self. At some level the ability to grant mercy, or forgive, is a repudiation of the sucker's framing.

Most families have their tales of alienation and forgiveness; mine is no different. My grandfather Luke was, in the language of the studies, "strongly averse" to compromise, and quick to take offense. When his fifth child, my mother, wanted to marry a non-Catholic man in a non-Church ceremony, there were tense negotiations, followed by impasse. My mother married my father in a field, no priest, no conversion; my grandfather disowned her. For years, she was not welcome in his home; he did not go where she would be. Once, their timing was slightly off, and he was leaving my aunt's house as my mother arrived. They met in the driveway, where my mom was holding her new baby: me. She tried to stop him, to show him his tenth grandchild. He walked past, silent.

Luke changed his mind when I was twelve, and he was eighty-two. He came to my cousin's wedding knowing we would be there. At the reception, my sister and I stood stiffly with our dessert plates as he introduced himself. He made a joke about having my cake and eating it, too, which I did not understand. After that, he started to come along when my grandmother would visit. He played chess with my father, his rediscovered son-in-law. We all got used to each other. Five or six years later, his health started to decline, and his children took turns taking care of him.

I asked my mother how she had come to accept him back into her life. Wasn't she furious? Didn't she feel taken advantage of? Luke had walked back into her life scot-free! He never even apologized!

She didn't see herself as any kind of fool for forgiving him. She hadn't compromised on anything that mattered. "When he came back, I had done everything I wanted. I got married and I had my kids and my whole life," she said. "I already had everything."

CHAPTER 4

Fight

Terse signs have begun to appear on American highways, instructing drivers: "MERGE LATE." If you're lucky, you get the unabridged version: "Use Both Lanes to Merge Point," although it's still a little unclear what mistake the signs are trying to correct. It turns out that merging is an unusually high-stakes engineering problem, and the instruction to "merge late" is the resolution of an internal debate among traffic planners, a coordinated reeducation effort to introduce drivers to the "zipper merge."

The zipper merge works the way it sounds: if two lanes are merging to one, cars stay in their lanes until the last minute, at which point the cars in the terminal lane move neatly into the through lane in an every-other-car pattern. The system only works if drivers approaching a lane closure actually postpone their lane shift until the last minute, rather than anticipating it and forming a longer, premature line in the through lane. If they queue up in advance, they have unwittingly sided with the "early merge" camp.

From an engineering standpoint, it's clear which system is more efficient: the zipper merge creates substantial time savings, because it maximizes the use of the road rather than lining cars up early and letting precious lane space lie fallow at the end. Unfortunately, humans, not just cars, are involved in merging, and as one traffic agency noted in response to the proposed changes, merging tops the list of "major auto-related causes of stress."

The journalist Paul Stenquist took the zipper merge dilemma to the streets, or at least the comments section, for the *New York Times*, posting a video instructional online and asking for readers to weigh in on the zipper merge. The response was discouraging to planning buffs but predictable to social psychologists: "Many said they would move into the through lane as soon as possible and were angered when others sped along until the last moment," he reported. "Some vowed that they would run off the road anyone who took this route. One respondent said the best argument against the zipper merge in the United States was that too many dangerous fools carried pistols and were willing to use them." Every merge point is a sucker's game, and in this game all the players are driving lethal weapons.

Where there is flight, of course, there is the looming possibility of fight, and the same themes that predict avoidance are not so much an impervious system as an unlit fuse; in the right conditions, the sucker's response flips from under-engagement to overreaction—revenge, retaliation, and even violence.

Imagine that you are in a line of traffic—frustrated, exhausted—and just as you are about to inch forward, a car zooms up on the right, from the breakdown lane. The car noses in front of you so that you can't really move forward without making contact, and you brake while the line jumper slides in. Can you feel your cortisol level rising? The feeling is so common that it now has a familiar name, "road rage," coined in the late 1980s to describe a purported rise in highway altercations.

At one level there is a primitive, instinctive response to being played for the fool. Driving in traffic is basically waiting in a slow-moving line, and people are primed to see each instance of turn taking or cooperation as an existential Public Goods Game. Indeed, if you google "road rage sucker," the search pulls up a series of local news reports of angry drivers exiting their cars to "sucker punch" each other.

The road rager who retaliates—refuses the merge, blocks the lane, tailgates in fury—mirrors the scorned cooperators from chapter 1, those Public Goods contributors yelling about alienation. Sometimes you can walk away from a scam. You can hang up or cross the street. But sometimes the mark is stuck behind a truck, and in that case things can turn primal. Everywhere from economics games to intimate violence to armed conflict, the "fight" shows up to retaliate, to insist on dominance or punishment, and to defend the social order from sucker threats.

Hitting Back

In their seminal paper on sugrophobia, Kathleen Vohs, Roy Baumeister, and Jason Chin puzzled over why the sucker's retaliatory instinct is so forceful. "Each year, several people die as a result of hostile physical encounters with vending machines," they wrote. "Many people feel angry when a vending machine fails to give the desired product, but for some people the anger apparently becomes intense enough that it leads them to shake the vending machine so fiercely to make it topple over onto them." More systematically, we have seen from experimental games that jilted players get really furious. They call each other "sons of bitches" and storm out of the lab—or, if they can fight back directly, they do.

The purest form of this, for me, is from an experiment called the Ultimatum Game. I first read the Ultimatum Game study in graduate school, in a paper that was, to be honest, almost unreadable. It was dense, jargony prose flanking dense algebraic tables. Once you got through that, though, the study setup was perfect. It was the kind of experiment that barely requires an explanation of the results, because once you see the design, you know what will happen.

The Ultimatum Game works like this. Players are paired up, and in every pair, one person is assigned to play the Proposer, and the other is assigned to play the Receiver. They have $10 to divide between them.* The Proposer is told to propose a division of the money, often sending a little paper form specifying the offer. Faced with this form, the Receiver has two choices. He can check "I accept" or "I refuse," sort of like a grade school crush negotiation ("Do you like me? Check yes or no"). If the Receiver checks "I accept," the players take the money according to the Proposer's division. So, for example, if the Proposer has suggested $6 for himself and $4 for the Receiver, and the Receiver accepts, they go home with their respective winnings. If the Receiver checks "I refuse," the players each take home nothing. That's the whole game.

The experimenters were economists, and economists typically predict that "rational agents" will try to maximize their own welfare. On this theory, a Receiver should accept any non-zero amount, since any money is better than no money.

This, of course, is not what happens. The Proposer has to suggest a split, and different splits have different appeal. Can I interest you in $5-$5? How about $9 for me, $1 for you? Most people can predict the outcome without ever reading an academic article. When the split looks too uneven, Receivers start to reject. Once you get down to $8-$2 or $9-$1, virtually none of them accepts.

Unlike the vending machines, the Ultimatum Game has never been fatal, but it does inspire predictable—and costly—revenge. Why would anyone turn down $1 or $2? It seems spiteful and self-

---

* I am again using the shorthand of $10 here. The original game was played in Germany in 1978, so they were playing with deutsch marks, and in fact the experimenters varied the "endowment," i.e., how much to give the Proposer from trial to trial. The principle of dividing and sharing a pie is, of course, consistent across endowments. Also, as in many of these games, the language that the experimenters use with the subjects is less descriptive and more mechanical than the description we use to explain the game: here, for example, the players would have heard that the roles were just called "Player 1" and "Player 2."

destructive to refuse money just so that someone else can't "win." The answer, of course, is that sometimes $2 is just money, and other times it's chump change.

An Ultimatum Receiver, offered her cut, only has two options: cooperate or detonate. Imagine a Receiver faced with a range of offers for his cut of the money. Would he take $5? Three dollars? One? We might be tempted to think he is looking at these offers and asking himself, *Would I like to have a little extra cash?*—but we know better. If that were the question, the answer would almost always be "Sure!" But it's not about whether I want a dollar; it's about whether I'll agree to a 10 percent cut—the same number but a different framing. What researchers have gleaned from years of studies is that the Receiver is asking himself something more like *Am I a sucker to take this deal?* The answer to that question predicts when he will accept and when he will fight back.

Over the years, across many labs, Proposers have tried every offer from $0 to (very rarely) all ten dollars. After thousands of iterations of the game, we know enough to predict how subjects will behave. Offers that look fair and equitable won't push sucker buttons; acceptance of $5-$5 and even $6-$4 are close to 100 percent. No suckers here! Almost half of all Receivers, though, will reject a $7-$3 offer, and over 80 percent will reject $8-$2. You can forget it with the $9-$1 offer.

Put differently, a lot of players will pay for the opportunity to retaliate. You think you can put me down? For just a couple of dollars, I can ruin the whole game for you.*

---

* It is tempting to think that the stakes are so low that the subjects just don't care—that their stance is basically just posturing without consequences—but I don't think that's right. The subjects who show up for these studies are students, and in many contexts they care about getting $2. When I was a graduate student, you could get a falafel lunch for that much money, or a bottle or two of Yuengling. More rigorously, the phenomenon has been replicated using larger sums. One study, conducted in the Indian state of Meghalaya, showed consistent rejection patterns whether the total endowment was 20 rupees, 200 rupees, or 2,000 rupees—in a region with a median wage of 100 rupees per day.

Running Hot

In a field characterized by extraordinarily dry writing, Ultimatum Game articles have titles that are practically tabloid headlines: "Unfairness, Anger, and Spite: Emotional Rejections of Ultimatum Offers" and "Emotions, Rejections, and Cooling Off in the Ultimatum Game." The players just get really mad. There is a reason that the mark has to be, in the language of Erving Goffman, "cooled out"—because he's so hot.

In fact, in some Ultimatum experiments, the players communicate their rage directly. Professors Erte Xiao and Daniel Houser ran a version of the game in which they permitted Receivers to send messages. One Receiver, replying to an eighty-twenty split, wrote to his Proposer: "Sorry, I'm a person too. When the cards are all in my hand, you should try to appease me instead of offend me. There was a 50/50 split. It couldn't have been easier. So, since you decided you are obviously better than I am. [*sic*] You get nothing. Enjoy it, I know I will." The indignant reply went directly to the heart of the sucker threat: *You think you're better than me?*

In responses like this, we can also see what the anger is doing. Anger is a canonically social emotion; it's sparked by a set of inferences about other people's intentions and motivations, and it's deployed in service of changing other people's thoughts or behaviors. Anger stems from the understanding that the $2 offer is not just context-free money; it has an offensive meaning, that "you decided you are obviously better than I am." Evolutionarily, anger has a social purpose: to intimidate free riders into taking the angry person's welfare more seriously. As one team of researchers argued, the function of anger is "to defend against exploitation and bargain for better treatment." A standard Ultimatum Game doesn't leave the players with much room for bargaining, but they can still leverage their fury by detonating the whole deal.

This same pattern crops up in scenarios where angry marks have the choice to be more deliberatively vengeful, rather than reactively spiteful. Lawyers in particular are very familiar with, not to mention enriched by, this dynamic. People who feel "screwed over" will litigate—even tiny claims, even at great personal cost—just to "make him pay." In 2002 the researchers Ernst Fehr and Simon Gächter, whose work on fairness has become the gold standard in the field, studied calculated revenge in the lab. In Fehr and Gächter's version, Public Goods Game players were permitted to punish selfish group members. The punishment was levied as a fine, but the only way you could impose the fine was if you paid a little yourself.

Imagine, for example, a player in a four-person Public Goods Game who saw that he and two other players had contributed generously and were going home with $6, while another player who had contributed nothing would be going home with $10. The game could just be over—live and let live, as it were. But in this version the suckered players could alternatively choose to pay out of their own pockets to reduce the winnings of the selfish player. In the Fehr and Gächter model, for each dollar a cooperator paid, the defector would be fined $3—so for a $2 investment any single player could eliminate the differential between himself and his scammer. More than half of the duped cooperators chose to spend money on just such a fine.

Indeed, retribution as a response to exploitative behavior is not just a way of punishing bad behavior on principle. In this context, it is arguably a way of undoing the sucker dynamic. When the reprisal equalizes payoffs for the cooperator and the defector, there is no longer an unfair discrepancy in their outcomes. To lean again on *Carrie*'s avenging protagonist: Carrie was suckered and humiliated, yes, but then she burned down her high school and her whole town, killing scores of people and ultimately herself. Amid that chaos was the thread of a more standard story: if the would-be sucker can extract costs from her tormentor(s), she is arguably no longer a sucker at all! Although Carrie sacrificed her own life, she was not getting the short end of the stick, since everyone else was also dead; the stick was only short ends at that point. A prototypical sucker is someone who lies down and takes it, so a vengeful response is itself disruptive to the insult. A punisher refuses the raw deal even when the alternative is mutually assured destruction.

If anger, retaliation, and the consequent resetting of the social order are predictable responses to a sucker's ploy, the predictable extreme is violent retaliation. On one end of the vengeful spectrum are people who appear to lose control at life's little scams—those who shout at retail workers or tailgate the guy who cut them off on the highway. On the other end is something much darker and more consequential: the furious scramble to reinstate the status hierarchy with violent consequences for women, immigrants, subjugated racial and ethnic groups, and gay and trans people. When the stakes are high—and status stakes are always high—retaliation can be devastating.

Margaret Atwood once wrote that she asked a male friend why men feel threatened by women:

> "I mean," I said, "men are bigger, most of the time, they can run faster, strangle better, and they have on the average a lot more money and power." "They're afraid women will laugh at them," he said. "Undercut their world view." Then I asked some women students in a quickie poetry seminar I was giving, "Why do women feel threatened by men?" "They're afraid of being killed," they said.

In fact, the fears have a logical order. Embarrassed men pose real threats to the women they think were laughing. Women lying about sex—who they're having it with, who they want to have it with, why they are having it with whom—is the archetypal female scam, and almost every culture has norms and narratives to constrain women's sexual choices. Violent revenge is a way to reestablish dominance, and the deepest threat to male dominance is women's sexual freedom; violent control of women defines and enforces the patriarchy.

Girls get told a lot of stories about the ways that men can't control themselves: the tank top makes the boys distracted; teasing will give your boyfriend "blue balls." And cheating will make him furious. When men discover their partner's infidelity, so the story goes, they are unable to control their jealous rage. The cuckold is naturally driven to avenge his dishonor.

And, indeed, accusations (founded or not) of female infidelity often precede intimate partner violence, up to and including homicide. The evolutionary psychologists Margo Wilson and Martin Daly spent years studying the phenomenon of uxoricide—wife killing—and observed, "There is a cross-cultural ubiquitous connection between men's sexual possessiveness and men's violence. The discovery of wifely infidelity is viewed as an exceptional provocation, likely to elicit a violent rage, both in societies where such a reaction is considered a reprehensible loss of control and in those where it is considered a praiseworthy redemption of honor."

The cuckold's revenge is not even reserved for actual cuckolds. In 2014, a college student named Elliot Rodger murdered six people in Santa Barbara, California, and injured thirteen more. His fury was not just at being rejected, it seemed, but at being lied to about the rules of the game. Women claim they want nice guys, but instead they give their attention, and sex, to "alphas." Rodger recorded a rant before his rampage: "I don't know why you girls aren't attracted to me. But I will punish you all for it," he screamed. ". . . I'm the perfect guy and yet you throw yourselves at all these obnoxious men instead of me—the supreme gentleman."

Rodgers had been "red pilled" by an online community of sexually frustrated men—"involuntary celibates" or incels—where he was led to the manufactured epiphany that men like him are merely marks for women who would, at most, use them for financial and emotional support. When they "realize" that women have been lying to them—lying about wanting love and kindness when in fact they want physical attractiveness—they start to fantasize openly about violence or take up actual arms. Rodger hoped to use violence to vindicate his status: "You are animals and I will slaughter you like animals. I will be a god . . ."

It is not just that violent men do, descriptively, respond to sucker threats or sucker harms with reactive aggression. Retaliation, especially in service of white male dominance, is socially sanctioned.

Wilson and Daly, who had described a pattern of domestic violence, also framed their findings in normative terms. Husbands across cultures do kill their wives, yes, but in part they are able to do so because their rage gets normative deference: "Indeed, such a rage is widely presumed to be so compelling as to mitigate the responsibility of even homicidal cuckolds." The anthropologist Jane Schneider observed that in "honor cultures" the family often engages in extraordinary vigilance over women's sexual perfidy, punishing girls who falsely claim virginity and wives who falsely claim chastity. Female sexuality in these contexts is sometimes described as inherently "treacherous," an inevitable tool for the exploitation of men.

It is tempting to see patriarchal policing of sexual freedom as something foreign, but the social acceptance of violence as a natural means of quashing the line jumper or the two-timer thrives close to home. Violence has always been understood as a natural male response to being taken for a mark. Our culture has empathetic understandings of a variety of violent behaviors, specifically when the violence comes from a white man who has been taken advantage of. Violence toward women who are cheats or even "teases" has a long and ignoble history. Violence toward Black men, especially, has often been justified by elaborate fears of disrespect or romantic entreaties toward white women. In the nineteenth century and into the modern era, that vengeful ethos has been encoded in the criminal law. A murderous response to adultery could be excused under the law, letting the killer off with a reduced penalty with what is called the "provocation defense."

The concept of provocation, legally and rhetorically, still does considerable cultural work to normalize violence, against trans women in particular. In 2003, when four teenage boys killed a trans girl named Gwen Araujo, their attorney argued to a jury that discovery of her deception was so provoking that they killed her in a heat of passion. The mother of one of the killers was quoted saying, "If you find out the beautiful woman you're with is really a man, it would make any man go crazy."

When defendants attempt the Trans Panic defense, they are leveraging the intuitive normative connection between deceit and retaliatory violence. Talia Mae Bettcher, a professor of philosophy at Cal State Los Angeles and a scholar of transgender studies, has argued that the stereotype that transgender people are "deceivers" is used as a baseline justification for violence against them. In her essay "Evil Deceivers and Make-Believers," she links the concept of sexual deception with a broader acceptance of violence against transgender women. She focuses on the rhetorical impact of linking trans gender identity with deceit: "I am specifically concerned with the ways in which victims of transphobic violence can be subject to blame shifting through accusations of deception." She argues that the rhetoric of deception—the idea that a trans woman can be "discovered" to be "really a boy" irrespective of her self-identification—sets up a story that feels familiar and plausible and encourages empathy for the aggressor. The grim plot has a temptingly logical progression, from exposure, to "accusations of deception and betrayal," to "extreme violence and finally murder."

Eric Stanley, a professor of gender and women's studies at the University of California, Berkeley, has observed, "Most forms of anti-trans violence are specifically brutal. They're also very corporal. Trans people are positioned in relation to a normative culture that is both fascinated and repelled by us. It's not usually, 'I hate you, get away.' It's more often, 'I hate you. Come really close so I can terrorize you.'"

Call to Arms

The rhetoric of the duplicitous "other" can be used to justify violence and also to incite it. From revolution to racial subjugation to ethnic cleansing, many demagogues have apprehended the electrifying force of a sucker narrative.

Perhaps we should not be surprised that our own recent former president, who leaned heavily on sucker rhetoric in all situations, ultimately used his incessant accusations of fraud and cheating to call for violent insurrection. After his electoral defeat in 2020, Donald Trump argued that Americans were being cheated by phony voters, the fake media, and weak or avaricious legislators—people who were conning them out of their rightful president. "When you catch someone in a fraud, you're allowed to go by very different rules," he told a crowd outside the White House on January 6, 2021. He was speaking specifically to the constitutionality of bringing certification back to the states, but he was sounding a familiar note: the situation warrants extralegal retaliation. Trump announced, "We will not be intimidated into accepting the hoaxes and the lies that we've been forced to believe. Over the past several weeks, we've amassed overwhelming evidence about a fake election. This is the presidential election. Last night [after the Georgia recount] was a little bit better because of the fact that we had a lot of eyes watching one specific state, but they cheated like hell anyway." As his followers looked toward the Capitol, he exhorted them to fight back: "You have to show strength, and you have to be strong." The subsequent attack was a violent takeover of the Capitol building, resulting in five deaths and dozens of injuries before it was brought under control.

How do you get people—specifically people who *could* hang back, who are not personally, emergently under threat—to support and even perpetrate atrocities? One way is to try to convince them that they are being taken advantage of, disrespected, or deceived. This was the logic of lynch mobs in the post-Reconstruction South. As W. E. B. Du Bois wrote, "How is it that men who want certain things done by brute force can so often depend upon the mob? Total depravity, human hate and *shadenfreude* [sic], do not explain fully the mob spirit in this land. Before the wide eyes of the mob is ever the Shape of Fear."

In Du Bois's view, their fear was that Black people would take their class status, their women, and their jobs. The way to inspire violence—to recruit for the mob—was to insist that Black citizens were the real breachers of the American social contract. One southern governor explained that lynching was, regrettably, sometimes necessary to prevent exploitation of white generosity: "Charitable and indulgent as we have ever been to an inferior race, cheerfully contributing bountifully of our time and means toward their material and moral betterment, still, if the brutal criminals of that race . . . lay unholy hands upon our fair daughters, nature is so riven and shocked that the dire compact produces a social cataclysm, often, in its terrific sweep far beyond the utmost counter efforts of all civil power." Claims of Black duplicity, from seduction of white women to labor organizing to plotting protest or insurrection, were rhetorical fodder for racial violence.

Adolf Hitler took a literal page from the book of white supremacists when he fomented a genocide against Jews in Europe, right down to threats that they would steal Gentile women. He convinced a nation to join him in protracted, decimating war and genocide on a tale of the humiliation of the Treaty of Versailles and the deviousness of Jewish citizens. His racist incantations accused Jews of a panoply of sucker's ploys, the destruction of Christian society enacted "not through honest warfare, but through lies and slander." Ultimately, his rhetoric directly echoed the language of American white supremacy, i.e., we have been generous, but they have taken advantage:

For hundreds of years Germany was good enough to receive these elements, although they possessed nothing except infectious political and physical diseases. What they possess today, they have by a very large extent gained at the cost of the less astute German nation by the most reprehensible manipulations.

He railed that the "special privileges" they enjoyed must be countered with, eventually, total annihilation.

Punishment

The reactive rage of a sucker is a primal response, a social practice, and a powerful weapon. In fact, it is so intuitive, and so normative, that it does not have to be reactive at all. People who are merely observing or judging social or economic quandaries are also re-markably invested in punishing schemers. The urge to punish is easy to defend in the case of serious crimes; it seems right that a society should penalize con artists who defraud the elderly or betray the public trust. But research consistently shows that when people perceive a scam—even if the scenario is ambiguous, even if the stakes are low, even if the punishment has no deterrent effect—their instinct is to make the schemer pay.

The *New York Times* used to run an economics advice column of sorts, staffed by authors and associates of the bestselling *Freakonomics* books. I followed the column, half professionally and half casually, because they often drew unusual connections between society and behavior. Some of the topics were downright psychological, including one particular query from a reader who offered this conundrum. It was 2009. He had bought his house at the market peak for a lot of money and taken out a substantial mortgage. Like many others who had done the same, his loan was now underwater, meaning he owed the bank much more than he could possibly get if he sold the house. He was in good company. As the country was reeling from the fallout of the 2008 financial crisis, one of the most pressing financial disasters for families and for the economy was home foreclosures. Millions of American homeowners had huge mortgages on homes with values that dropped below the loan balance: people were paying off $400,000 loans on houses that wouldn't sell for half that price.

Generally speaking, if you stop paying your mortgage, your home goes into foreclosure, meaning the bank can repossess it and the borrower in default is evicted with a black mark against their credit score for the next seven years. In most cases, foreclosure is a last resort, a worst-case scenario for both the homeowners and the bank. But the lax lending standards, the housing bubble and crash, and the rapid readjustment of other market indicators—including, crucially, the stabilization of the residential rental market—led to an unusual situation in which foreclosure was an affirmatively attractive option for some borrowers.

In the case of the letter writer, he still had his job and was otherwise able to pay his bill every month to the mortgage company. But he had done some math and realized that there was a cheaper option. Because he lived in a "no-recourse" state—a state in which mortgage lenders cannot sue borrowers for other losses after foreclosure—it would be cheaper, all told, for him to call the bank, tell them he was giving them the keys to his home, and just . . . walk away. This was true even with the hit to his credit score and even accounting for the challenge of finding new housing.

Most people who default on a loan do so because they are out of options; they're stuck and they default because they can't pay. By contrast, the letter writer was looking at a different kind of dilemma, the prospect of a "strategic default": he would default by choice, not necessity, to improve his financial picture. There was even a website linked to this option, YouWalkAway.com, where you could enter in your financial details, including credit information and local housing costs, to see how much money it would save you to strategically default. So this reader was asking: What should he do?

The column was open to public responses, and they quickly showed a stark split. Some people looked at the problem and thought: Your mortgage is a kind of an option contract where your choices are to either (a) make your loan payments to the bank or (b) give your house to the bank. Commenters who brought that frame to the problem gave practical advice about the order of financing and how to manage poor credit. Other people, though, saw the situation as a scam, a con that the homeowner was running on the bank and possibly the whole economy. Their responses were less sanguine. One such commenter wrote, "You cut a deal with another party. You need to live up to that deal . . . Forget credit scores. Do you want to be a cheat and a liar?" Some of these readers thought that he should be punished, by the bank or by the states, for breach of the mortgage contract, which they saw as a kind of fraud.

Or, more vividly: "Since you're willing to sacrifice integrity for money, why not just rent out your wife for a few years and pay off the whole thing very quickly?"

Now, the writer in question was raising a problem with his debt repayment in an era in which mortgage lenders had arguably kicked off a global recession. Predatory lending, a practice in which lenders would offer unnecessarily bad terms to unsuspecting borrowers, was the scam at the heart of a generation's macroeconomic reckoning. Nonetheless, the commenters who saw the strategic default as a hustle were irate—even though it didn't affect them personally and even though the choice was a foreseeable risk for the banks and a financial benefit for the homeowners.

I was at first taken aback by this discourse, because it seemed especially vicious to me: Why would you bring his wife into it??? But I recognized the vibe, of course, in part because of my own experience posing contracts dilemmas to people with strong intuitive responses. I was really curious to understand the crux of the fury: What was it about breach of contract that was so offensive? What feeling was it evoking? To study the question systematically, I showed subjects different transactional harms: floors that did not get refinished, a yard that did not get landscaped, a party that did not get catered. For each harm, I randomly assigned subjects one of two conditions. They read either that the loss was the result of (a) an intentional breach of contract, or (b) negligence by a stranger. I asked: How much should the wrongdoer have to pay in damages?

Subjects consistently reported that the breach of contract should be more seriously punished than the negligent harm, even though the financial loss was the same across scenarios.

In these contracts scenarios, people saw scams and wanted to punish the perpetrators. With a colleague, Dave Hoffman, I dug further into the emotional substrate of their responses. A pattern emerged: punishment was associated with a feeling of disrespect and betrayal. What predicted the punishment response was the extent to which the subjects saw the breach as a status hit. Punishment is more than just reactive lashing out; it is a direct response to social derogation. There was nothing special about the status relationship between the floor refinisher and the person whose floors he didn't refinish, but when he tried to pull a fast one, it looked like a play for dominance, and suddenly the status stakes were salient.

The political philosopher Jean Hampton once wrote that punishment has the effect of reinstating the hierarchy. Taking advantage of someone else is literally a power play—it lowers the mark and bolsters the operator—and punishment is a way to knock the wrongdoer back down the ladder: "The one who acted as if he were the lord of the victim is humbled to show that he isn't lord after all," wrote Hampton. "In this way, the demeaning message implicit in his action is denied. Therefore, just as the crime has symbolic meaning, so too does the punishment. The crime represents the victim as demeaned relative to the wrongdoer; the punishment 'takes back' the demeaning message."

Kenworthey Bilz, a law professor and social psychologist at the University of Illinois, took this as a jumping-off point for an experimental test of responses to a criminal fraud. She showed subjects a con game—in this case, identity theft—and asked them to rank the relative social standing of the victim and the offender. In one condition, they were told that the perpetrator would not be punished for targeting the victim. In the other condition, they read that the offender had been punished for victimizing the person whose credit she stole.

Subjects were asked to rate, on a scale, the social standing of the victim and of the offender. Without punishment, people perceived the offender as having a higher community standing than the victim. When the offender was punished, though, subjects regarded the victim as higher in status than the criminal. As Jean Hampton had said, "[T]he punishment 'takes back' the demeaning message."

Whose Scam Is It Anyway?

Punishment is a way of fighting back against the con, at scale. If one of the functions of punishment is the reinstatement of hierarchy, and if exploitation is especially degrading, then the pattern I have been describing is predictable: There is a strong drive to punish free riders. But some of the examples we have seen of free riders drawing ire should give us pause. Why, for example, are people up in arms about individual homeowners pulling a fast one on corporations? In some of these cases, the strength of the response is not proportionate to the extent of the harm—the banks were just fine and indeed handily bailed out by the federal government—but rather to the disruptive effect on the status quo. Corporations leverage their power over individuals all the time, but it can feel uniquely destabilizing when individuals try to do the same in return. More importantly, the appeal of harsh punishment for perceived scams is not confined to the bloodless world of mortgage boilerplate. When people who are traditionally relegated to low status—women, people of color, immigrants—breach their contracts, the punitive response can be especially drastic.

In my contracts course, I teach a vivid historical example of this phenomenon, a case called *Bailey v. State of Alabama.* To understand the context of the case, it helps to keep in mind the connection between breach of contract and disrespect, and the insistence on Black respect for white supremacy in the post-Reconstruction American South. Given these premises alone, we can almost predict to the letter how southern states reacted to Black employment mobility after slavery. In the South, Black freedom was a threat, and it was treated legally as a presumptive fraud on white people.

Nowhere was this more explicit than in the 1911 breach of contract case of Lonzo Bailey. Bailey was a Black man working as a sharecropper. He had agreed to a one-year contract to work on a cotton farm but left before the term was up. His employer sued. In this case, they did not just sue for breach of contract, which would have been draconian enough; they pursued criminal penalties under a bizarre Alabama law of fraud. The case went to the U.S. Supreme Court, which ultimately found for Bailey—not because the court didn't think he had breached his contract, but because of the extraordinary lengths that the state of Alabama had gone to try to punish disloyal sharecroppers.

We have every reason to believe that the terms of Bailey's employment deal were exploitative, and those exploitative terms were, of course, set on a backdrop of racial hierarchy enforced with real and threatened violence. Sharecroppers received clothing, food, and sometimes housing, the costs of which were charged against future wages. Payment was deferred to the end of the agricultural cycle, and it was tied to the market price of cotton rather than a fixed hourly wage. The accounting was often based on the word of the landowner. Irrespective of the truth of the matter, the boss could not be productively or safely accused of lying by Black laborers, because accusations would be taken as rank disrespect. Rampant fraud perpetrated on laborers was the norm.

Lonzo Bailey's employment contract was not a good one; it was a bad deal offered to people who were largely excluded from the labor market's better opportunities. Many people would want to quit that job, which was perhaps why the Alabama legislature was moved in 1896 to pass section 4730, a law that said that any person who received money or property from an employer and subsequently left without repayment was criminally guilty of intentional fraud.

For contracts lawyers, this is an astonishing statement. The fundamental move of contracts is to make it possible for private citizens to do business with one another—even risky business—without making them vulnerable to state punishment. Fraud applies to people who intentionally deceive at the outset. Fraud can be criminally punished; George C. Parker was committing a crime when he sold and resold the Brooklyn Bridge. But if I say that I am going to deliver apples on Monday and then, come Monday, I am all out of apples or my delivery truck breaks down or *even if I just change my mind*—that's a breach of contract, which is not a crime. If I don't sell my apples as promised, I can be sued in court for the losses I have caused, but that's it. I have not defrauded anyone; I cannot be criminally prosecuted.

What the state of Alabama did, which the Supreme Court rejected fifteen years later, was to announce in its statute that anyone who quit their sharecropping job—the statute didn't refer to sharecroppers but it was clearly passed with them in mind—would be *assumed* to have had the intention to defraud. That is, breach of contract itself would be taken as evidence that the employee intended to fraudulently extract benefits, like clothes or cash advances, and abscond with them. The law would make breach of the employment contract punishable by hard prison labor. (The Supreme Court said the statute was a violation of the Thirteenth Amendment's abolition of slavery and involuntary servitude, and the statute was struck down.)

It was a vicious, and literal, policing of the exploited as exploiters. But in a system of white supremacy, it was crucial to close off Bailey's avenues for social mobility and punish even the most inchoate attempts to transcend the racial caste. Coming down hard on minor schemers, however marginal or marginalized, was one way to do this.

The Supreme Court invalidated the Alabama law in 1911, but the pattern remains familiar over a century later. Even our tax system reflects some of these pathologies of weaponized vigilance, extravagantly policing low-level scams while struggling to articulate the principles that would permit intervention when the perpetrators are among the powerful.

Take, for example, fraud by individuals on the government via the tax system. In many ways, the American tax system is an honor system; underenforcement is built in. The primary verification mechanism is the IRS audit. The IRS cannot or does not confirm the accuracy of every filing, so instead it chooses a sample and relies on a combination of social or moral norms with the threat of a positive probability of audit. In 2018 the nonprofit news organization ProPublica published a story on the distribution of audits in America. One of the most frequent triggers of an audit is receipt of the earned-income tax credit.

People who claim the EITC are receiving a government benefit; it is a massive anti-poverty program. Dorothy Brown, the tax scholar and law professor at Georgetown University, specifically identified the aggressive EITC audit strategy as a response to fears of welfare cheating. She wrote:

The error rate on EITC claims remains high because the credit continues to be very complex. The solution is for Congress to *simplify* the EITC, not for the IRS to audit more people. The fact that the government has taken the opposite approach is itself revealing. If you believe errors are inadvertent, then you reduce the complexity of the process. But if you believe errors are intentional "welfare cheating," then you increase the rate of audits.

Small-time scammers who write bad checks, steal from petty cash, fudge their residency details, or use food stamps for non-food items—these are harms the legal system understands and is prepared to name and punish.

There is a recursive pattern to accusations of scamming. One party realizes what looked like a fair transaction was actually a raw deal; the accused exploiter retorts, "You think I'm scamming you? *You're* scamming *me*!" Americans who have been cheating undocumented workers out of their wages for decades (centuries?), for example, might soften the blow of accusations by claiming that, in fact, undocumented workers are stealing American jobs! Especially when both parties are acting against background conditions of economic unfairness, they can both make some claims to truth.

The legal theorist Aditi Bagchi, a professor at Fordham Law School, wrote a provocative article motivated by the following example: a street vendor selling a scarf claims that it is 100 percent cashmere, but it is just polyester. Bagchi stipulated that the society in which the buyer and seller both live has failed to live up to "its obligations of distributive justice." She posited that the vendor's poverty and the buyer's wealth are the direct results of the unfair system in which they both live. On these facts, she argued, the vendor might be lying, but he is not cheating. He is engaged in civil disobedience, protesting the underlying exploitative economic conditions.

Whether I am inclined to agree with her argument in theory, I feel confident that in practice people are much more inclined to see the scam in a scarf sale than in an economic system. Where there is room for interpretation, the energy goes to protecting the status quo. Retaliatory aggression and violence in reaction to minor scams is both normal and normalized. Getting angry and fighting back when you feel suckered is a response that is easy to relate to. We know from anonymous games and contrived studies that people approve of punishing free riders. But in the context of an actual society, it is important to track who, exactly, is eliciting this violent response. In 1896 the state of Alabama was not about to legislate more humane working conditions for exploited Black farmworkers, but it did pass a law to protect the landowners from being "defrauded" by sharecroppers. Especially in high-stakes conflicts with underlying power imbalances, the question is not whether the marks fight back but rather which marks get to claim victim status.

Aggression against schemers is expected and even valorized, and there is an appetite for narratives that reinforce the social order. High-status exploiters may deflect the anger otherwise directed at them by claiming that their low-status victims are the real operators. Some of the most successful claims of victimhood seem to be reserved for those who have the most status in the first place. In the meantime, people with limited political power requesting redress for demonstrable exploitation find themselves in an upside-down world where some imagined minor perfidy on their part is a bigger deal than real systems of oppression.

We blame the victim when it feels better than blaming the real aggressor. A truthful story about the exploitative status quo feels really disruptive, and that feeling can be dissipated by refocusing the retaliatory instinct on a more palatable villain. Sometimes someone who is expected to accept subordination makes a claim: I have been exploited, this is not fair, I am the victim here. The perverse consequence of that claim is that it creates incentives to find a less disruptive narrative, an alternative accusation that will bring that hostility down on the natural losers and let the winners be.

CHAPTER 5

Racial and Ethnic Stereotyping

The very construction of racial and ethnic categories is built with the raw materials of sucker fears. In some contexts the connection between racial narratives and fools' fables is abstract or allegorical, but when it comes to contracts, it's literal—which is why I am starting a chapter on race with a story about contract law and getting it in writing.

There is a school of thought in contracts research called relational contract theory (bear with me). It grew out of an intellectual movement at the University of Wisconsin in the 1960s, and its genesis was an ethnography of Milwaukee businessmen who submitted to a series of sociological interviews about their professional practices. The resulting research literature was one of the first truly empirical accounts of contracts outside of litigation—who makes them, what they say, what happens when they fall apart—and it was viewed as a fairly radical critique by a left flank of the legal academy.

The surprising nub of the results focused on the relationship between the research subjects and the legal system. The men who were interviewed reported candidly that they found their formal contracts—their carefully negotiated and expensively lawyered agreements—distasteful, even irrelevant. What really mattered, they said, was picking up the phone and calling the other guy involved in the deal, not your lawyer. You stake your reputation on the firm handshake, not the fine print. And when things go wrong, you lean on goodwill and reputational consequences, not courts or money damages; if you want to do business with your buddy again, you have every incentive to make the deal work rather than suing over every mishap.

The relational contract theorists seemed to think that the future of contracts was the increasing obsolescence of formality: no more legalese, just reasonable fellows getting so-and-so on the horn. For its time, it was a progressive battle cry of contract theory.

There was a key fact to this account that got little explicit attention but came through on a close reading. "The top executives of the two firms may know each other," wrote one researcher. "They may sit together on government or trade committees. They may know each other socially and even belong to the same country club."

The country club! Now we know one more thing about these trust-based business arrangements: they were for white men.

About fifteen years later, a young lawyer named Patricia Williams left her position as deputy city attorney for the city of Los Angeles and started her first job as a law professor. Williams was one of very few Black faculty members in law schools in the early 1980s. Her fields, contracts and commercial law, were even less diverse than the professoriate as a whole.

Like the relational contract scholars from Wisconsin, she, too, was interested in contract formalities. But after a decade in legal academia, she wrote her most famous essay on that topic not from her perspective as an attorney but from her vantage as a consumer. Her work intervened at a time in which most of the people using complex contracts, practicing commercial law, and teaching contract law would have consistently experienced the legal world from the same white, male starting point. The essay, "The Pain of Word Bondage," was published in 1991 in her book *The Alchemy of Race and Rights*, and it upended a century of assumptions.

Williams was herself an attorney and an elite academic, an Ivy League graduate (Harvard Law School) who ultimately became an Ivy League law professor (Columbia Law School), a published author, and a MacArthur Genius fellow. "The Pain of Word Bondage" described how she, a Black woman, and her colleague Peter, a white man, had each started working at Columbia the same semester and each had to rent an apartment in New York City. Here is the story, in her words:

> It turned out that Peter had handed over a $900 deposit in cash, with no lease, no exchange of keys, and no receipt, to strangers with whom he had no ties other than a few moments of pleasant conversation. He said he didn't need to sign a lease because it imposed too much formality. The handshake and the good vibes were for him indicators of trust more binding than a form contract. At the time I told Peter he was mad, but his faith paid off. His sublessors showed up at the appointed time, keys in hand, to welcome him in. There was absolutely nothing in my experience to prepare me for such a happy ending. (In fact I remain convinced that, even if I were of a mind to trust a lessor with this degree of informality, things would not have worked out so successfully for me: many Manhattan lessors would not have trusted a Black person enough to let me in the door in the first place, paperwork, references, and credit check notwithstanding.)
>
> I, meanwhile, had friends who found me an apartment in a building they owned. In my rush to show good faith and trustworthiness, I signed a detailed, lengthily negotiated, finely printed lease firmly establishing me as the ideal arm's-length transactor.

While her colleague Peter bumbled affably along, Williams, by her account, measured her steps along a tightrope. She perceived peril on either side; she might be seen as overly credulous or insufficiently trustworthy. The documentation and the formality protected her from both. Peter used informality to reinforce his relationships—the willingness to be trusting signaling that he was neither suspicious nor suspect.

Williams observed that the social meaning and the social value of trust differs by race and status. Were she to behave as Peter had done, gamely exposing herself to exploitation, it would not have the same meaning or the same results. Her essay echoed themes that recur for people of color, reminiscent of Barack Obama being hounded for his birth certificate or the Black residents of Ferguson, Missouri, resorting to videorecording to prove their claims of police brutality. People who expect to be under suspicion know the exhausting importance of documentation. "I was raised to be acutely conscious of the likelihood that no matter what degree of professional I am, people will greet and dismiss my black femaleness as unreliable, untrustworthy, hostile, angry, powerless, irrational, and probably destitute," Williams wrote. "Informality in most white-on-black situations signals distrust, not trust. Unlike Peter, I am still engaged in a struggle to set up transactions at arm's length, as legitimately commercial, and to portray myself as a bargainer of separate worth, distinct power, sufficient *rights* to manipulate commerce."

Williams knew that in the American cultural landscape her very participation in contracting was suspect. Mulling this example with me, a friend cited the widely held view among Black shoppers of the importance of documentation. The attorney Lawrence Otis Graham wrote in the *Washington Post* that, when their son was a teenager, he and his wife kept a list of rules for navigating a racist world. Number 4: "Never leave a shop without a receipt, no matter how small the purchase, so that you can't be accused unfairly of theft." Some people are presumed to be suckers, and others schemers, and others maybe both, not because of their behavior but because of their ethnic or racial identity.

The American status quo is racial hierarchy, and the rhetoric of that hierarchy is entwined, even rooted, in narratives of marks and hustlers. Consider, for example, the consistent just-so stories of racism, the insistence that people of other races are foolish, devious, or both. Even the language of racial stratification, down to the specifics of ethnic slurs, invokes the fear of a destabilizing scam at a remarkably literal level.

To be blunt: sucker rhetoric is deployed—variously, creatively, relentlessly—in service of white dominance.

Social Dominance and Racial Stereotyping

The late Harvard psychologist Jim Sidanius was raised in Manhattan and the South Bronx and then left the United States for Sweden to complete his graduate work. He, too, was a Black scholar in a predominantly white field of study. In an interview with the *Harvard Crimson* in 2017, Sidanius was asked how he came to his life's work, his theory of oppression and intergroup relations. "One of the things I noticed while I was in Sweden was that even though the level of prejudice and discrimination in the culture was much less than I experienced in the U.S., it was still the case that there seemed to always be a group of people who were discriminated against, and oppressed society would always divide into dominants and subordinates," he responded. "In reading through history it seems to be the case that it was a constant phenomenon in all societies regardless if they were democratic or authoritarian. They all consisted of systems of group-based inequality, and it was an attempt to try to understand why that was."

Sidanius and his colleague Felicia Pratto published the definitive social science account of intergroup oppression in the book *Social Dominance*. They observed that there is a universality to certain kinds of social stratifications. Every known culture has age-based and sex-based hierarchies; male dominance over women and adult control of children is near-universal. In addition, crucially, almost every culture has what Sidanius and Pratto

termed an "arbitrary-set system." The arbitrary-set system strati-
fies members of a society by socially constructed and salient char-
acteristics like ethnicity, class, religious sect, clan, nation, race,
"or any other socially relevant group distinction that the human
imagination is capable of constructing." Sidanius's insight was
that there is nothing special about the criteria societies use for
stratification—they have no inherent meaning—but every society chooses among some set of easy differentiators, creates a system of dominance and subordination, and then dedicates resources
to preserving that order.

The power of the social dominance theory is that it aims to explain a wide range of human behavior in terms of the ultimate goal
of social dominance. On this account, racism has a goal, and the
goal is power.

Racial and ethnic stereotypes do the work of social dominance,
providing "legitimizing myths" to justify the arbitrary hierarchy,
according to Sidanius and Pratto: "Most if not all forms of group
prejudices, stereotypes, ideologies of group superiority and inferiority, and forms of individual and institutional discrimination both
help produce and are reflections of this group-based social hierarchy." The stories a culture tells about different groups "provide
moral and intellectual justification" for social inequality: *they* don't
value hard work; *they* pulled themselves up by the bootstraps; *they*
deserved it.

The legitimizing myths of racial hierarchy—of white dominance, specifically—often cast ethnic and racial minorities in the
caricatured roles of scammer and mark. For an efficient exposition
of stereotypes and ideologies, a culture's vicious humor can be extraordinarily revealing, as in a bigoted old joke about three men
greeting St. Peter in heaven. The setup devolves quickly into racist
caricature, the Jewish man scheming to defraud a credulous Black
man while the white Episcopalian looks beatifically on. In joke
form, the racist disdain and the poisonous anti-Semitism add up to
a menace all the more pernicious for its faux joviality. Maybe even
more importantly, the one-two punch of bigotry is not just a set of

insults; it is also a sly argument against interracial solidarity. Stereotypes and prejudice are not just raw animus, or an up or down vote on a whole race or ethnic group. They have content, and the content does the work of justifying who gets what in a society.

Nearly contemporaneous with the social dominance model was a parallel approach to the psychology of stereotypes from a group of researchers at Princeton. Although they came from a different perspective, their theory echoed the insights of Sidanius. Professors Amy Cuddy, Susan Fiske, and Peter Glick were interested in what beliefs constitute intergroup prejudice. A lot of research that talked about racial bias or sexism would describe those attitudes as basically being one question: How much animus do you have toward this group?

Cuddy, Fiske, and Glick had observed that this kind of one-dimensional measurement failed to capture the real cognitive processes in stereotyping. Groups are stereotyped in specific ways, not just good or bad but "good musicians" or "bad athletes" or, more to our point, "untrustworthy" or "easily misled."

The researchers were asking how stereotypes get constructed, and they proposed a theory. When groups of humans meet each other, they suggested, they have to quickly determine two things (part of the idea here is to imagine groups meeting each other on the plains or a desert, pre-civilization, presumably worried about physical survival):

1. Are these people friendly or antagonistic toward me?
2. Can they execute their goals competently?

Cuddy, Fiske, and Glick argued that stereotypes are formed out of the answers to those questions; they called the theory the stereotype content model. They predicted that most groups could be plotted in this two-dimensional space, along axes labeled "warmth" (friendly versus antagonistic) and "competence."

To see the logic, it can be helpful to think about examples at the individual level. A mentor, for example, would be high on both dimensions. A sniper from an enemy army—high competence, presumably, but certainly low warmth. Your own toddler? Off-the-charts warmth, not so much competence. The idea was that these same dimensions are relevant to group stereotypes. The way we respond to people from a particular group can be predicted by where we place them on the grid.

The key insight is that stereotypes are stories, not just free-floating animus. Those stories—the "content" of the stereotypes, or the legitimizing myths of the social hierarchy—motivate a predictable set of behaviors. The theory has often been summarized in a table like this:

	Low Competence	High Competence
High Warmth	Paternalistic Prejudice Elderly, housewives *Pity, sympathy*	Admiration In-group and close allies *Pride, admiration*
Low Warmth	Contemptuous Prejudice Poor people *Contempt, disgust, resentment*	Envious Prejudice Rich people, feminists *Envy, jealousy*

It can be really jarring to see stereotypes mapped out this way: stereotypes are rarely laid so bare and, plotted bloodlessly like this on a graph, they feel extra pernicious. (Good. They are.)

On this model, prejudice starts with a stereotype, or a cognitive appraisal. This is a belief system about the attributes of the group. That stereotype yields an emotional response: affection, for example, or disgust, or envy. Emotions are uniquely motivating, and the affective response translates the cognition into action: cooperate, thwart, aggress, share, help, discriminate. This is how the stereotype motivates behavior.

The theory gives us a new tool for thinking about the role of sucker rhetoric in group stereotypes. It might be a warning: Watch out, they want your stuff; watch out, they're tricky. Or it might be a justification: Poor fools, this is what's best for them. Or it might be both: This is why you have to be vigilant; this is why you can't let them in.

Indeed, we can see the sucker mapping even more clearly if we think about synonyms for "warmth" and "competence." What if we replaced "competence" with a range of adjectives from "foolish" to "savvy," and "warmth" with a spectrum of "trustworthy"? The meaning stays the same, but the sucker stakes become much clearer:

	Foolish	Savvy
Trustworthy	Suckers	Paragons
Untrustworthy	Parasites	Schemers

There is an in-group, often white men; everyone else is defined, at least implicitly, in relationship to that reference group: the pitied suckers, the feared schemers, and the parasites held in anxious contempt. Gridded out in this way, we can see clearly some of the legitimizing myths of our own social order. The clichés, jokes, fables, and just-so stories of the American melting pot yield a cheat sheet of bigotry, an executive summary of whom we are meant to disdain, and whom we are meant to fear.

Ethnic stereotypes that depict whole groups as foolish or naïve, for example, abound. If you've ever heard the apocryphal story of the purchase of the island of Manhattan from the Lenape Indians, you have heard such a myth. American schoolchildren used to hear the story of the Dutch merchant who bought Manhattan: "Glittering beads and baubles, brightly colored cloths, glittering trinkets of small value brought from the ships nearby in chests, and opened on the shore before the eager eyes of the aborigines, were what worked the miracle," recounted one history from the early twentieth century. The island was purportedly sold for $24 in beads that "filled the minds of the simple Indians with delight."

People from Central and South America have been painted with a similar brush, accused in childish myths of handing over a civilization when they were fooled into believing that the Spanish explorer Hernán Cortés was actually the Incan god Quetzalcóatl, a constructed history that fit neatly into the theory of the Indigenous population as naïve and gullible. In a through line to the present, anti-Latino prejudice has long included stereotypes meant to justify subjugation on the basis of incompetence: lack of education, or intelligence, or work ethic.

Indeed, racial stereotyping so depends on sucker narratives that we use ethnic identifiers as literal shorthand for hustles. I still periodically hear a student in contracts class unthinkingly deploy an ethnic slur, hopefully having no sense of its etymology, when they read a case and protest that one of the parties has been "gypped." Discrimination against the Roma (referred to as "Gypsies" originally in a mistaken shorthand for Egyptians, which they are not) is longstanding and ongoing, and one of the principle stereotypes is duplicitousness. A now-outdated dictionary gave an example sentence, "Look out for this guy; he's a clever agent to slip you a gyp," and indicated that the word was "derived from the popular experience with thieving Gypsies." The Roma, who migrated west from northern India over the course of a millennium, are cast in a longstanding trope of the migrant as swindler. Even in the relatively anodyne Wisconsin sociology of contracts, one of the researchers wrote that a core tenet of relational contracts was that "one does not welsh on a deal." The xenophobic verb "welsh" in fact refers to England's western neighbors from Wales, who were apparently accused of reneging on racetrack bets.

And, perhaps most saliently, American Jews are the eponym for a sucker-related slur: to "jew down" means to negotiate insistently, to insist on a bargain. Anti-Semitism is perpetrated by spreading the fear of Jewish scheming, from Shylock to the *Protocols of the Elders of Zion* to Holocaust denial. Anti-Semitic accusations have been part of American culture since long before the founding; in the 1650s, Peter Stuyvesant even tried to ban Jews from New York on the grounds that they were "deceitful."

The content of anti-Semitic bigotry is so firmly established that it can be pulled, as a trigger, to make people feel they are being scammed. As one team of psychologists wrote, "Agreeing that Jews are 'extraordinarily clever' is at least as likely to indicate dangerously anti-Semitic prejudice as the lack of it." In America, this is the rhetorical function of invoking George Soros. When millions of Americans protested the police murder of George Floyd in the summer of 2020, social media saw a spike in references to Soros and his funding of the protestors. Soros, a Hungarian-American investor and philanthropist, is Jewish; he has also been accused of funding migrant caravans from Central America to the southern U.S. border. In both cases, the ethnic prejudice leverages anti-Semitism not just at the purported target but also at other groups. Protestors who were reacting to the horrifying incidence of lethal police violence against Black men have a serious and legitimate claim—but the evoking of Soros was meant to make the compassionate response to those claims seem foolish. Maybe this outpouring of anguish is the exaggerated lies of paid protestors; maybe your sympathy means you have been duped.

Amy Cuddy, coauthor of the stereotype content model, wrote for the *New York Times* about the string of anti-Semitic attacks culminating in the Tree of Life synagogue massacre in 2018:

> When times get tough, envious prejudices can ignite. Societal breakdown, harsh economies or political turmoil can activate resentment toward high-status minorities, who are seen as competitors for limited resources or even [as] dangerous enemies. The stereotyped competence of such groups, when they are suspected or accused of "cold" ill will, suddenly represents a serious threat. Envy toward these groups becomes volatile, mutating into anger and spurring the most extreme forms of discrimination—intentional harm and even annihilation.

The same connection could be drawn between anti–Asian American prejudice and the 2021 mid-pandemic shootings at three Atlanta nail salons, where six of eight murder victims were women of Asian descent. Historically, white prejudice against Asians features stereotypes like "sly" and "insular"—that is, capable of duplicity and with loyalties that lie elsewhere. A research team at Princeton—Monica Lin, Virginia Kwan, Anna Cheung, and Susan Fiske—brought those dormant intuitions to concrete items on a scale to measure stereotypic beliefs. They found that the core tenets of anti-Asian bigotry were that Asian Americans seek status and economic success that they do not need or deserve. On this view, they "enjoy disproportionate economic success," "think they are smarter than everyone else," and "aim to achieve too much." These are serious accusations of line jumping and status hogging—the kinds of sucker invocations that historically lead to white violence.

White Americans are often resistant to the idea that widespread bigotry against Asian Americans exists at all—they are a "model minority" whose success is sometimes leveraged as evidence that white people are not racist—but we have seen many devastating reminders that this kind of ambivalent stereotyping has pernicious outcomes, too. Asian Americans have been literally accused of deception en masse: after Japan attacked Pearl Harbor in 1941, Japanese Americans were interned to prevent them from acting as spies or saboteurs. The coronavirus pandemic brought yet more reminders of white suspicion of Asian intentions, spurring a spike in anti-Asian harassment.

The Tightrope

Of course, the division I have proposed between fools and schemers is not so neat; even patronized groups are held in suspicion. When Donald Trump warned Americans that Mexico wasn't "sending their best," this was precisely his claim: Mexican migrants are trying to make you feel sympathetic for their plight, but actually they're taking your jobs and your women.

This duality is especially clear, and especially trenchant, in anti-

Black racism. From the first days of chattel slavery, justifications of violent subjugation espoused a belief that enslaved Black people were poor fools, incapable of caring for themselves, while insisting on constant vigilance against the threat of revolt or escape. The whiplash is the point. Any observation can be made into evidence of inferiority. Any data point can be leveraged for racial animus— each success evidence of conniving; each hardship evidence of foolishness or malingering.

There is a through line of racist tropes from construction of the enslaved person as willing fool (remember Kanye West: "When you hear about slavery for 400 years: For 400 years, that sounds like a choice") to the scientific racism of the twentieth century. A long and ignominious tradition of caricatures plays on stereotypes of Black fools: *Gone with the Wind*'s Mammy, Stepin Fetchit, and Uncle Tom. The culinary historian Michael W. Twitty has described Aunt Jemima as propaganda of Black consent to subjugation, "an invitation to white people to indulge in a fantasy of enslaved people—and by extension, all of Black America—as submissive, self-effacing, loyal, pacified and pacifying."

Anti-Black racism leverages the sucker construct to argue that Black people are fools while simultaneously, furiously, insisting that any racial progress is the illegitimate result of cheating, line cutting, and theft. American culture has a complicated relationship with the notion of Black deviousness, in part because of a long-standing commitment to the myth of Black intellectual inferiority.

But one look at our institutions suggests that Black Americans are treated as being deeply untrustworthy. Black communities are surveilled and policed with a relentlessness that serves as a constant reminder that they are perceived as an unacceptable threat. Often institutional actors—business leaders or government officials— offer pretextual cover to deny that they are targeting Black people, but sometimes the mask slips and we get a blunt admission of deep, ingrained distrust. In 2016, for example, the governor of Maine felt emboldened to make that racist subtext into text. Warning Mainers about Black men, he ranted, "These are guys by the name

D-Money, Smoothie, Shifty. . . . They come up here, they sell their heroin, then they go back home. Incidentally, half the time they impregnate a young, white girl before they leave."

For white Americans afraid of being suckers, the specific terror of losing out to Black people feeds into a vicious cycle of fear, derogation, surveillance, and retaliation. When it comes to who gets policed for everyday "frauds"—who is routinely accused of, and punished for, trying to put one over on the system—the most salient examples are Black people, and especially Black women. The story often has a recognizable pattern: A Black woman is subject to a set of policies best understood as structural racism. She pursues a goal that is fundamentally prosocial, caring for family members, community members, or fellow citizens. Then the same structures that caused her material disadvantage in the first place include a rule that accuses her of fraud when she does the exact work that she is otherwise accused of not doing.

Who is suspected of voter fraud? Crystal Mason, a Black woman on supervised release for a conviction for federal tax fraud submitted a provisional ballot, unclear about whether her conviction or sentence meant she could not vote. The vote was not counted; as a convicted felon on supervised release, she was not an eligible voter under Texas law. The claim that she committed voter fraud rested on a theory of Mason as an intentional cheater. The judge found that her signature on the provisional-ballot affidavit was dispositive: she knew she was voting illegally. The fact that the text in question was on the back of the paper she signed, and that she was not required to read it before voting, was not regarded as important. She was sentenced to five years in prison for intentionally defrauding the voting system.

Who is suspected of school placement fraud? Tanya McDowell, who was homeless, used the address of a friend to enroll her five-year-old son in a promising elementary school. She was convicted of felony larceny—that is, she was convicted of stealing state resources. Single mothers, and single Black mothers in particular, are subject to rivers of criticism for allegedly inadequate parenting. Black people are criticized relentlessly for a supposed lack of commitment to education. But when Black women try to obtain the very goal they are often accused of neglecting high-quality educational opportunities for their children—they go from being pitiable to suspicious. "Indeed, rather than seeing poor Black parents as stealing education, we could characterize their behavior as attempting to reclaim the education debt owed to them after hundreds of years of educational neglect," observed the law professor LaToya Baldwin Clark.

Who is suspected of fraud in sports? Serena Williams, a vocal anti-doping advocate, nonetheless wondered about the discriminatory practice of her own drug-testing regimen; she was tested at rates unheard-of for other athletes. She made a public complaint in 2018 that she was being bizarrely and intrusively targeted. Caster Semenya, the world's fastest female middle-distance runner, has been repeatedly accused of "gender fraud." She has been subject to testosterone testing and hormone-altering drug regimens in order to bring her natural hormones closer in line with the profile of other female runners.

Even the racist cliché of laziness is a subtle accusation of fraudulent intent. In the United States, white slavers and white landowners historically deployed the rhetoric of Black laziness to justify ownership, physical and economic violence, and white supremacy. Especially in the context of race stereotypes, "lazy" is an accusation of free riding—and free riding is always about suckers. The accusation was: This worker claims to be working as hard as they can but in fact they are holding effort in reserve and ultimately loafing on your dime. And, of course, accusations of laziness are now and have historically been deeply cynical, deployed to justify coercive moral correction and to ignore principled resistance.

Patterns we have seen in other chapters—resistance to redistributive policies that provide social assistance to poor people of color, violent retaliation to sharecropper resistance—are additional evidence that the stakes are not the same for white people and Black people navigating perceptions of exploitation. On the

one hand, when Black people have been persecuted or discriminated against, they are disdained as willing dupes to minimize the
urgency of their claims. On the other hand, they are cast as objects
of suspicion, ready to fool tenderhearted white people and take
their stuff (their money, their women, their status). It is a perilous
line to tread.

❖

Who gets to step off the tightrope? Recall the story about white
people in Wisconsin making contracts with each other: the deals
were easy, self-enforcing, embraced in a circle of trust and benefit
of the doubt. That warm feeling of contractual joviality arose for
me recently when I read a *New York Times* report on another amicable deal. A Danish museum paid an artist about $83,000 in cash
on a contract to make a pair of works that would somehow display
the cash "reflecting the nature of work in the modern world." The
artist, Jenns Haaning, returned two blank canvases to the museum, a piece of art he titled *Take the Money and Run*. He kept the
cash for himself.

The museum director reacted with good cheer. "The work is
interesting to me," he said. "It is partly a humorous comment: why
do we work, what is satisfying about being good at something?"

Contracts to purchase art are high-risk. There are no guarantees that a work of art will hold value over time, and there are often
risks of actual forgery. It is a Wild West in terms of markets, which
is one reason it is so ripe for money laundering. In fact, selling art is
risky, too: for successful artists, the buyer's (unenforceable) promises to steward the work are crucial to the sale. A buyer who turns
around and resells art at auction at a discount can affect the market
price of an artist's whole body of work. Overall, it is the kind of
unregulated, high-stakes, unstable market where we would expect
a lot of mutual suspicion. Jenns Haaning was openly devious. His
swindle was met with chuckles and equanimity.

I was reminded of a story that the poet Cathy Park Hong recounted in her essay collection, *Minor Feelings*. Hong had been on the edges of the New York art world in her twenties and was going to the solo show of a friend, a "guy named Joe," an artist who also played in a band. His friends grumbled that he had waited until the last minute to mount his show, and indeed, when Hong arrived, she found the walls "practically empty except for a few dirty un-primed canvases." It was visually irritating, and she expected others to react as she had. It turned out to be wildly popular, and Joe's career was launched. Ivanka Trump hung one of his canvases in her penthouse. Musing on the problem of what makes art *art*—not just a child's scribbles, not a pile of junk, not a canvas you forgot to prime—Cathy Park Hong observed:

> The avant-garde genealogy could be tracked through stories of bad-boy white artists who "got away with it," beginning with Duchamp signing a urinal and calling it art. . . . The problem is that history has to recognize the artist's transgressions as "art," which is then dependent on the artist's access to power. A female artist rarely "gets away with it." A black artist rarely "gets away with it."

Hong zeroes in on that age-old fear—What if this "art" is just junk and I'm the chump?—and sees the anxiety subside when the players are white guys: "The bad-boy artist can do whatever he wants because of who he is." What—me worry?

Hong's prediction that a Black artist rarely "gets away with it" has not been empirically tested in the art world, but evidence from other transactional categories suggests she is right. Where white counterparties might have a wide margin of error for one another—what's $83,000 between friends?—a series of case studies and experiments show a different picture for interracial transactions, a rigid refusal to risk getting the short end of the stick in a deal with a Black person. Many Black homeowners, for example, have recounted the importance of concealing family photographs and memorabilia in order to sell their homes, and even found that they get higher appraisal figures when they have a white person handle the transaction.

Professors Ian Ayres, Mahzarin Banaji, and Christine Jolls tested the effects of seller race experimentally in a clever study on eBay. They advertised a series of baseball cards, picturing the cards being held by either a light-skinned hand or a dark-skinned hand. The cards were sold to the highest bidders, per the eBay auction rules (a timed auction rather than a "Buy It Now" transaction); those held by the dark-skinned hand sold for about 20 percent less.

This same pattern of skepticism holds when Black people enter the market as buyers. Many researchers have pointed out that Black homeowners pay more to own their homes, having been offered worse mortgage terms and higher insurance premiums. The Yale Law professor Ayres devised another experiment, this one a controlled study to test the cost of being a Black buyer. He hired thirty-eight different people in four demographic categories— white men, white women, Black men, and Black women—to go around to ninety Chicago car dealerships and negotiate the best deals they could for new cars. In each negotiation they recorded the salesperson's initial offer and their final offer. The difference between the white men and the Black buyers was significant: Black men, most notably, were charged over $1,100 more than white men for the same car, final offer.

These results are overdetermined; that is, they are explained by multiple plausible causes that likely apply simultaneously. But they strike me as evidence of a risk aversion when it comes to the imagined potential of being exploited, however marginally, by a Black person. It echoes the results of the Bohnet and Zeckhauser Trust Game from chapter 3—the game where betrayal-averse subjects would gamble with a computer but not with a human. Here, the risk of feeling scammed by a Black counterparty is so aversive, white sellers would forgo a so-so sale that they would have permitted with a white buyer. Like Alabama's response to Lonzo Bailey—every breach an affront—the insult would be too serious a violation to permit.

Special Favors

Until 1833, there were criminal consequences for people who borrowed money and didn't pay it back. Those who were truly insolvent would be sent to debtor's prison, subject to state penalties for private breaches. The abolition of debtor's prison was one of the reasons that the Alabama state legislature had to jump through such jurisprudential hoops to find a way to hold breaching sharecroppers criminally liable. Today, people who borrow money and don't repay it—whether the loan is from a bank or a car dealership or even friends—can argue to a court that they can't pay it back and, therefore, they should be eligible to file for bankruptcy.

If you have to file for bankruptcy, and you are filing as a person rather than as a business, there are two ways to do it: Chapter 7 and Chapter 13, and the choice is up to the debtor. The differences between these can be a little obscure, but they matter to the person filing. If you file for bankruptcy under Chapter 7, you pay low fees and you get a onetime reckoning. Your credit score is wrecked, but the debt is discharged and you can move on with rebuilding your life.

Chapter 13 has the same negative impact on the debtor's credit score, but it costs more up front and takes longer to resolve. It has higher attorney's fees and administrative costs, and it puts the filer in a supervised relationship for three to five years after she files. In a Chapter 13 filing, the court typically appoints a trustee who ensures that, during the period of supervision, all of the debtor's disposable income is garnished and distributed to the waiting creditors. As you can deduce from these descriptions, for most people Chapter 7 is the most sensible choice. It is cheaper and faster. The main reason to file under Chapter 13 is to avoid foreclosure on a home, because Chapter 13 permits long-term restructuring of mortgage debt.

Bankruptcy has always been controversial. From one perspective, it appears to let reckless or irresponsible debtors take advantage, put one over. Filers are often suspected of being "won't-pays" rather than "can't-pays," a classically subtle accusation of suckerdom aimed at those who might think that debt release is a compassionate policy. When Elizabeth Warren was still a law professor and not yet a senator, she was part of a consortium of bankruptcy scholars who started to ask questions about how bankruptcy actually happens in the United States. One of her most influential findings was that most debt is not "irresponsible" spending; it's medical debt.

In 2012, that same bankruptcy research group (now sans Warren) took notice of something anomalous in their combing of national bankruptcy data: rates of filing under Chapter 13 specifically seemed to vary dramatically by region, with its peak popularity in the southern states. Moreover, Black debtors appeared much more likely to have their debts discharged under Chapter 13 than white debtors. Three researchers, Jean Braucher, Dov Cohen, and Robert Lawless, decided to investigate the race discrepancy systematically. First, they verified their anecdotal impression: White, Latino, and Asian American households were choosing to file in Chapter 13 about one-quarter of the time. African American filers chose Chapter 13 over Chapter 7 a whopping 55 percent of the time.

Braucher, Cohen, and Lawless were perplexed. We would expect Chapter 13 filers, as a group, to have especially high rates of homeownership, since that's the main reason for using Chapter 13. But statistically, Black Americans are much less likely than whites to own a house, yet much more likely than whites to file in Chapter 13.

The research team decided to send an experimental study to a sample of bankruptcy attorneys to see whether lawyers were steering their clients by race. Each attorney received a set of facts about a fictional couple, including a list of their assets as well as their outstanding debts. Half of the attorneys read about Todd and Allison, who belonged to the First United Methodist Church, and the other half read about Reggie and Latisha, who belonged to the Bethel AME Church. (One additional control group also read the story with initials only and no race signifiers.) Overall, 32 percent of the attorneys thought that Todd and Allison should file in Chapter 13, but when it was Reggie and Latisha, the rate went up to 47 percent.

The attorneys had no idea they were counseling clients based on race; when asked if they thought there was a different approach for Black and white filers, none thought the researchers would find anything. But we know from their actual responses that they implicitly drew a connection between race and bankruptcy chapter.

When you file for bankruptcy, you are breaching a bundle of contracts in an orderly and court-supervised fashion. Promises to pay your credit card company, your internet provider, unpaid utilities, and medical bills are all broken. You can see the bankruptcy system as a compassionate intervention for the desperate, or a socially efficient way to get people back into the economy, or maybe a pathetic sucker's game. Indeed, when the bankruptcy laws were reformed in 2005, Kansas congressman Todd Tiahrt argued that the changes were necessary to prevent people from using bankruptcy "as a tool for fraud to cheat their way out of debt." Chapter 13 is one way of putting the brakes on breach forgiveness. Who gets away with it? Not Black filers.

The Texas law professor and bankruptcy scholar Mechele Dickerson has homed in on the pervasively moralistic reasoning in bankruptcy debates. Bankruptcy reform was necessary because "Congress also felt that it was crucial to restore 'personal responsibility and integrity in the bankruptcy system' and prevent abusive, morally bankrupt debtors from opportunistically discharging their debts," and debtors seeking discharge in Chapter 7 were "abusive, opportunistic, and strategic actors." Discharging debts in bankruptcy is efficient and humane, a kind of systemic mercy. But it is a favor that even Black filers' own attorneys seem reluctant to bestow on them.

❖

We commonly understand a "favor" to mean an act of kindness beyond what is due. By definition, a favor is an undeserved benefit, and if someone gets "special favors," the undeserved benefit is presumptively illegitimate. The kinds of schemers who *get* special favors are usually the diplomat's kid, the teacher's pet, or the boss's mistress. The people *accused of asking for* "special favors" are often people of color, enforcing their rights or insisting on equal treatment.

Special-favors rhetoric is at the core of racial resentment, an observation that the sociologists Emmitt Riley and Clarissa Peterson made systematic. Racial resentment is constituted by the beliefs that "Blacks should try harder, Blacks are no longer the subject of discrimination, Blacks should work their way up without any special favors, and Blacks have already received undeserved advantages." Resentment is bitterness that comes with being treated unfairly; these statements suggest that white Americans believe that they are being treated unfairly by Black Americans. They suggest a belief that Black people claiming entitlements based on past discrimination are overstating their case and the "special favors" are unfair; they are duping white people out of their stuff and their status. This leads to the perverse view that when Black people make claims for equality, they are actually cutting in line. Insisting on equal rights becomes evidence of taking advantage of the system. For psychologists, the content of anti-Black racism centers around exploitation and deceit, the belief that African Americans are actually exaggerating their reports of discrimination and harm.

In psychological studies of racism, subjects often take a short questionnaire to indicate their agreement with statements about racial prejudice that have been tested and validated in the population. Looking at the standard instrument for measuring racism, the "Modern Racism Scale," a picture becomes clear: anti-Black racism is a posture of deep skepticism. Those who score highly on the scale agree that anti-Black discrimination "is no longer a problem in the United States" and endorse a series of complaints about Black Americans, including that they are "getting too demanding in their push for equal rights" and "have gotten more economically than they deserve."

The special-favors discourse creates a deep challenge for the goal of racial progress, because it makes it seem like the stated goal, equality, is actually a Trojan horse, a cover for the hidden aim of white subordination. An unequal society stays unequal in part by sowing fear of the destabilizing scam, insisting racial progress and other claims to equality are actually schemes to steal status.

Psychologists Michael Norton and Samuel Sommers, faculty members at Harvard and Tufts, respectively, studied the relationship between racial attitudes and perceptions of anti-white discrimination. They asked white and Black respondents to rate the seriousness of racial bias by decade. For each decade, subjects had to indicate on a 10-point scale the overall seriousness of anti-Black bias and the overall seriousness of anti-white bias. Black respondents indicated that anti-white bias has been minimal the whole time, a mostly flat line at the bottom of the graph, and anti-Black bias went from very high to moderate (over 9 on a 10-point scale in the 1950s to 6 in the new millennium), a marked downward slope. By contrast, white respondents' ratings of anti-white bias was almost a perfect mirror of their ratings of anti-Black bias—that is, they thought it was low in the 1950s but inched up over time. For every tick down in discrimination against Black people, they saw an uptick in discrimination against white people. They rated contemporary anti-white discrimination as *a more serious problem* than anti-Black discrimination. "Not only do Whites think more progress has been made toward equality than do Blacks, but Whites also now believe that this progress is linked to a new inequality— at their expense," they concluded. The title of the article? "Whites See Racial Progress as a Zero-Sum Game."

CHAPTER 6

Sexism and Suckerdom

J udge William C. Pierce was five years from retirement when he adjudicated his most famous case, a claim of consumer fraud and misrepresentation. Judge Pierce had started his professional life at fourteen, leaving school to work on a rail yard; he got his law degree by taking night classes at the Atlanta Law School. He worked as a lawyer for the Department of Agriculture for thirty years before he was appointed to the state court of appeals in Florida, where he was clearly exasperated to have to resolve a dispute between a housewife, Mrs. Vokes, and her dance teacher, Mr. Davenport.

The matter of *Vokes v. Arthur Murray, Inc.* preserved his contempt in amber; all scare quotes are from the original judicial opinion:

Plaintiff Mrs. Audrey E. Vokes, a widow of 51 years and without family, had a yen to be "an accomplished dancer" with the hopes of finding "new interest in life." So, on February 10, 1961, a dubious fate, with the assist of a motivated acquaintance, procured her to attend a "dance party" at Davenport's "School of Dancing" where she whiled away the pleasant hours, sometimes in a private room, absorbing his accomplished sales technique, during which her grace and poise were elaborated upon and her rosy future as "an excellent dancer" was painted for her in vivid and glowing colors. As an incident to this interlude, he sold her eight ½-hour dance lessons to be utilized within one calendar month therefrom, for the sum of $14.50 cash in hand paid, obviously a baited "come on."

Thus she embarked upon an almost endless pursuit of the terpsichorean art during which, over a period of less than sixteen months, she was sold fourteen "dance courses" totaling in the aggregate 2,302 hours of dancing lessons for a total cash outlay of $31,090.45, all at Davenport's dance emporium.

In the end, Audrey Vokes had paid $31,000—adjusted for inflation, that's over $250,000!—for thousands of hours of functionally unusable dance lessons.

It turned out that in order to climb the Arthur Murray Dance Studio rankings, you had to purchase specialized packages of lessons, which then triggered eligibility for different forms of clubs, conventions, and status designations. Buying lessons you'd never use was sort of like buying airline miles just to hit Platinum, only with less concrete rewards. Vokes's claim to the court was that she had been defrauded; knowing she was terrible, but wanting her to buy extra lessons, Davenport had told her that she showed dancing promise. Hers was an unusual claim, because the scam was more subtle and more personal than a typical breach of contract. It wasn't that the studio had promised her lessons and canceled them, or promised that lifetime members would be eligible for particular exhibitions or reservations and reneged; it was that Davenport had sold Vokes lessons with cynical flattery, falsely claiming he had a positive opinion of her dancing. In fact, it was widely agreed among the instructors that she could barely find the beat, much less dance to it. (Do you want to dance? No, I said you look fat in those pants!)

Audrey Vokes was a woman without a man, a job, or a bullshit detector. Judge Pierce could barely bring himself to even take the idea of *dancing* seriously—much less paying money to learn about dancing, much less being a fifty-one-year-old woman with a new hobby. Her new pursuit was a "yen," and her diligent engagement was just whiling away the pointless hours. I don't even know what's going on with the part about her "absorbing his accomplished sales technique" in a private room—fine, yes, I do know, and it's gross—and it's very hard to grasp what jurisprudential work was being done when Judge Pierce wrote, "In other words, while she first exulted that she was entering the 'spring of her life,' she finally was awakened to the fact there was 'spring' neither in her life nor in her feet."

Judge Pierce could hardly have been more condescending, or derisive, which is why his decision is particularly strange. The law was not exactly on Audrey Vokes's side; her only real claim was that Davenport had lied about his opinion of her. This kind of lie usually doesn't count as misrepresentation in the legal sense.* If I go to a store and try on a dress and the sales clerk says, "It looks great!" I can't go back and cancel the transaction later because I wore the dress and found out it just looked so-so.

Judge Pierce did not have to rule for Vokes, a woman he claimed to disdain, but he did. He called her foolish, but he ruled that she could be protected from predatory businesses like the dance studio.

Reconsidering the reasoning of the *Vokes* opinion, I was reminded of a false advertising suit from the 1940s when the Federal

---

* Normally parties to a contract can rely on representations of fact. If I say my house has no termites and in fact it has termites, the buyers who believed me can back out of the deal, because I have misrepresented an important fact and they relied on my statement when they agreed to the deal. By contrast, if I tell them my house is great for entertaining, or in a really fun neighborhood, those are just my opinions, and the buyers would not be able to rescind their offer based on the claim that they relied on my opinions when they made the deal.

Trade Commission made a women's cosmetics brand change the name of their face cream. The cream was called "Rejuvenescence," and the court ruled that it was too misleading; the name would fool women into thinking that their skin would literally get younger. The court wrote:

> The law was not made for the protection of experts, but for the public—that vast multitude which includes the ignorant, the unthinking and the credulous . . . [like] the average woman, conditioned by talk in magazines and over the radio of vitamins, hormones, and God knows what.

I have not taken expensive, fruitless dance lessons, but I have certainly purchased expensive, fruitless cosmetics. Like most women, I have run through a gauntlet of scams that play out in the same bewildering form: You should try to be appealing to men; this product or service promises that outcome; actually you are defective and so it doesn't work for you; you were very foolish to spend money on this—what were you thinking??? Face cream, dance lessons, low-rise jeans, hot yoga, Special K cereal. The Arthur Murray "come-on" was unusually elaborate but not unusual in its structure.

When I lived in New York City in my early twenties, I was, like many women in my demographic, subject to two recurring street hustles: (1) vouchers for a facial and massage at a nearby spa, and (2) the invitation to an "open call" for modeling tryouts. I don't know what happened if you went to the modeling tryouts, because I was not delusional. I believe the scam involved prepaying for headshots or something like that. However, once, lonely, having resisted for weeks, I did take the spa voucher, talked into it by a person yelling compliments at women on the street.

For some reason the voucher had to be used immediately, which I dutifully did. Once you got to the "spa"—it was in fact a nail salon—you got ten minutes of a punishing, clumsy facial and then the rest of the hour was just a hard sell for "add-ons": "half a head of highlights" or a French manicure. When I asked for just the package advertised on the voucher instead, the aesthetician regarded me with raw contempt. (At one point, frantic to repel the sales pitch, I brandished my fingers, bitten to the quick since grade school, like *I* was outwitting *them*—like, Ha! You couldn't manicure these if you tried!) I ended up paying extra anyway, tipping the workers and sheepishly buying some sort of exfoliating scrub.

When I crossed the street on my way out and looked back at the salon, I saw the scene afresh. The recruiters were just teenagers, reckless and insouciant with stacks of misspelled flyers, yelling bald-faced flattery at paralegals and editorial assistants like me. I had moved to New York optimistically and held on to a fragile, aspirational sense of myself. As a minor scam, the spa located me socially and with devastating accuracy: a human resources associate at a midtown accounting firm, dressed from the Banana Republic sales rack, who did not have other plans that night.

For me, as perhaps for Audrey Vokes, there was a heteronormative, physical quality to the con—compliments, touching—that brings to mind a comment by two esteemed social psychologists, Laurie Rudman and Peter Glick: "A strong sign that gender, sexuality, and status are intertwined is illustrated by the use of sexual euphemism (i.e., 'getting screwed') for being robbed or duped. The person 'getting screwed' is, metaphorically speaking, being equated with the female role during sex."

This is a provocative view, the idea that the ones who are taken in, convinced, betrayed, and duped are tautologically *feminine*, no matter whether the particular chump in question identifies as a man or a woman. Men and women can both play the fool, but something about being suckered is womanly. In turn, being accused of a feminine foible has very different implications, depending on the target: it is a rank insult when aimed at men (think of a military commander or an athletic coach cajoling an all-male team, "Let's go, ladies!") and it is something more ambivalent when directed toward women. You can't understand suckers without grappling with gender; more controversially, you can't understand sexism if you don't reckon with sugrophobia.

The Gendered Sucker Schema

Psychologists, especially those who focus on studying stereotyping and gender roles, have systematically unpacked the relationship between gender norms and sucker norms. The psychology of gender has a deep history and a wide scope (and I'm getting there, I promise), but I want to start with a small, recent study. In 2014, psychology professors Laura Kray, Jessica Kennedy, and Alex Van Zant wanted to understand a single specific stereotypic belief: that women are easy to mislead. They were inspired by a 1981 self-help manual for car buyers, apparently written by a veteran salesman who was full of cautionary tales. Indeed, the salesman had a whole sucker folk theory, which he called the "typically uninformed buyer": "As a rule they were indecisive, wary, impulsive, and, as a result, were easily misled. Now take a guess as to which gender of the species placed at the top of this 'typically easy to mislead' category? You guessed it—women."

Kray, Kennedy, and Van Zant took this "typically uninformed lady buyer" theory seriously and tested it experimentally. They recruited participants to read a short story about a car buyer who has responded to an ad for a used car. Half of the participants were randomly assigned to read that the buyer was Michael, a "typical male negotiator." The other half read the same scenario, but now starring Patricia, a "typical female negotiator." Subjects answered some questions about this buyer: How kind will they be? How high in business sense or ambition? How gullible or naïve? Are they expected to be "easily misled"? The results were clear. Patricia was rated as being significantly more easily misled than Michael.

What does it matter if someone thinks you are more or less gullible? The research team had a second hypothesis: if people think women are easier to fool, then savvy negotiators—not even necessarily sexist, just playing the odds—will be more likely to try to mislead women than men.

They tested this prediction with data from a business school course where the students had been paired up and asked to complete a simulation negotiation over the sale of a house. The exercise had been originally conducted as part of an educational module on ethics in negotiation. Students had been given a hypothetical dilemma with a complex ethical hurdle embedded in it: a seller was offering land with the hard stipulation that it could only be sold to someone who would use it for personal residential purposes, but in fact the buyer wanted the land to build a high-rise hotel. Students role-played the negotiation in pairs. Both sides knew the seller's condition, but only the buyers knew the secret high-rise plan.

The students in the buyer role had an incentive to lie. The exercise rewarded pairs who closed their deal, and the buyer would be more likely to get the seller's consent with a promise not to build. That would be false, of course, but there was no formal penalty for lying.

Kray, Kennedy, and Van Zant wanted to know who got told the truth and who didn't. When men were in the role of seller, they were deceived about 5 percent of the time. When women were selling, though, the buyer would lie in 25 percent of the deals—five times more often. Women: easily misled, and therefore actually misled. In the language of the stereotype content model you'll recall from the last chapter, they had been pegged—warm, maybe, but not competent.

Both men and women can play the sucker, and both men and women can run the con—but they are not playing the same games when they do. The fool is feminine, or feminized, in the shared imagination. The sucker construct is inexorably gendered, and that gendering has predictable consequences. When Daisy Buchanan announced in *The Great Gatsby* that "the best thing a girl can be in this world [is] a beautiful little fool," she had a take on what it means for a woman to be easily misled. Women who take the bait or fall for the come-on are not admirable per se, but they are playing a part that has been written for them. Most scams are at the structural level in any case—no paid family leave, pink taxes, the wage gap—and we have to deal with them whether we're subjectively fooled or not.

When I teach the psychology of gender, I remind students that, as a scientific field, it is changing in real time. Accurate descriptions of sex stereotyping in the 1970s felt off by the new millennium; the research I studied myself in college in the 1990s already looks outdated now. Specific scorn around "career women" or even "female promiscuity" does not resonate with Zoomers the way that it did with Generation X. But there is a through line in gender stereotyping even as the particular content shifts over time: gender norms are sucker narratives—who is allowed to put one over, who is likely to be misled, who takes it lying down. Damaging gender norms dictate that a duplicitous woman is especially transgressive, raising the terrifying threat of cuckoldry. The other side of the same coin is that it's more acceptable, or at least predictable, for women to play the fool. From inferior intellects to the impostor syndrome, the legitimizing myths of male dominance hold on to a theory of women as cooperative dupes.

Of course, the *very* early psychological theories of gender were pretty straightforward: women have small brains, which is why they are subordinate to men. Discounting the more dubious history of phrenology, penis envy, and hysteria, the first systematic social psychological accounts of gender stereotypes were published in the 1970s. To modern readers, they will feel like a snapshot of their time, a turning point when second-wave feminism was spurring scientific progress, but it was also still cute to call a woman "highly suggestible."

Sandra Bem, the pioneer of this field, finished her doctoral thesis in 1968, right as the world was changing, and published the Bem Sex-Role Inventory (BSRI) soon after. Bem became well-known first for her research, then for her unconventional marriage and parenting philosophy, and ultimately for her death, a suicide carefully planned to come before the worst ravages of Alzheimer's disease.

Many people still know her best for the psychological instrument that bears her name, which she debuted in an article called "The Measurement of Psychological Androgyny." Bem was interested in people like herself who did not conform to standard expectations of women. She wanted to develop a scale that would accurately measure contemporary social mores of gender and also permit a respondent to score high on both masculine and feminine traits—"*both* assertive and yielding, *both* instrumental and expressive," in her words.

There is a methodology to studying stereotypes. The research goal for psychologists like Bem is to understand the distinctive content of stereotypical beliefs. We saw this in the chapter on racial stereotyping, where the items on the Modern Racism Scale were specifically about resentment and special favors, not just animus in general. Similarly, in the context of sex stereotyping, the psychological instruments like the BSRI look to disentangle what can feel like a mishmash of prejudicial beliefs. Bem was interested in "sex-typing": she wanted to know what was meant, specifically, by masculine or feminine roles in a society.

To understand what it meant to people to be feminine or masculine, she aimed to differentiate positive personality traits. Her insight was that everyone aspires to positive personality traits—people want to think of themselves as smart, and kind, and loyal—but that we can learn something from the distinct intensity of each aspiration for men versus women.

Bem and her research assistants compiled two hundred adjectives that they thought sounded both admirable and also gender-specific. They also tested two hundred additional traits that they guessed would be gender-neutral.* One hundred Stanford

---

* The non-gendered items included things like "moody," "conceited," "unsystematic," and "sincere." Even those words seem from my vantage to be associated with one gender—I would have guessed feminine, masculine, feminine, feminine, but that was not statistically borne out by Bem.

undergraduates, half men and half women, each took the list of four hundred total traits and rated every trait. For each adjective, the question was the same: "In American society, how desirable is it for a man/woman to be _____?"

Bem then used the responses from that survey to whittle the items down to make a manageable but coherent psychological instrument. For each sex-typed category, she chose items that were rated as significantly more desirable for one sex than the other, by both male and female judges. Among the traits that met those criteria, she chose only twenty for each category. The resulting instrument is sixty items long: twenty masculine items, twenty feminine items, and twenty gender-neutral items. (You can take the BSRI online, like a BuzzFeed quiz, and it purports to tell you whether you are feminine, masculine, or androgynous. There is also a concerning category called "undifferentiated" for people who basically fail to identify with any of the sixty listed traits.)

What the Bem Sex-Role Inventory offers, for our purposes, is two lists of ostensibly positive gender-specific adjectives, i.e., this is what it means to be good at being a man; this is what it means to be good at being a woman. Here is a sample:

Masculine:	Feminine:
Analytical	Childlike
Individualistic	Flatterable
Competitive	Tender
Forceful	Gullible
Dominant	Affectionate
Athletic	Yielding

These items, which describe social attitudes toward masculine and feminine norms, suggest a specific relationship between men, women, and the hustle. For men, the sucker construct is not complicated, or at least not complicated by gender roles. Being a sucker is about weakness: men should dominate; therefore being a sucker is a gender role transgression. Psychologists who have written about patriarchy and social power have sometimes described the purpose of sex stereotyping as "hegemonic masculinity," a theory whose core proposition is that the point of masculine norms is to defend male status—to defend the status quo of male dominance.

The picture is more puzzling when it comes to women. It is not that women are supposed to be suckers, exactly—but is it really true that "affectionate" and "yielding" are aspirational traits? In the late 1990s, twenty years after Bem debuted her research, psychologists looked at the BSRI and raised an objection. The Bem traits were all chosen specifically because they were viewed as *positive* attributes. But the feminine adjectives in particular seemed to suggest a confusing model of femininity: a person who is flatterable, gullible, and yielding is going to get duped. Certainly in the new millennium it would be unusual to use "childlike" as a compliment for an adult. Are grown women supposed to *aspire* to a state of yielding gullibility? Did the original scale capture something robust? Did it accurately speak to phenomena that endured over time?

Professor Deborah Prentice and a doctoral student at the time, Erica Carranza, proposed an update to the BSRI that was not about changing the traits but instead focused on changing the structure. In their view, it's not that women *ought to be* gullible, but that gullibility can have positive connotations for women (Eliza Doolittle, Snow White, most Zooey Deschanel characters) that it rarely does for men. The survey that they conducted was written up in a 2002 paper called "What Men and Women Should Be, Shouldn't Be, Are Allowed to Be, and Don't Have to Be," and the theory was true to the title. Prentice and Carranza found that gender norms are a web of mandates and loopholes that work together to define the boundaries of social success and acceptable deviance on the one hand, and true norm violations on the other. There are also gender norms about which negative traits are more acceptable for women to show than men, and vice versa, or which good traits are more acceptable to lack. And there were also strong proscriptions: there were gendered third-rail character flaws. I've organized a sample of the results into tables to highlight the implicit sucker pattern:

Women Must Be	But Cannot Be	And It's OK if They Are	And It's OK if They Are Not
Friendly	Intimidating	Naïve	Worldly
Compassionate	Cynical	Weak	Rational
Patient	Stubborn	Impressionable	Competitive

Men Must Be	And It's OK if They Are	And It's OK if They Are Not	But They Cannot Be
Assertive	Stubborn	Helpful	Weak
Rational	Self-righteous	Cooperative	Gullible
Self-reliant	Controlling	Sensitive	Yielding

To be blunt: women are expected to be up for playing the sucker; men are supposed to avoid it at all costs. Women must be patient and compassionate; men must be assertive and rational. No one wants to be a sucker, but if a woman is naïve, it's all right. Most people don't want to be seen as domineering, but if a man doesn't want to cooperate, that's okay.

For women, playing the sucker is the cost of doing business; you can't be cheerful and cooperative all the time without getting the short end of the stick here and there. At the everyday level, there are little dilemmas, minor occasions that raise the pressure to play along—the LuLaRoe invitation, the insincere compliment, the spa coupon—that stand in for the existential con. On a job interview early in my career, for example, I met a senior male faculty member who introduced himself to me as Abraham Lincoln. I still have no idea what the purpose of this joke was, or if it was even a joke per se. I remember thinking, *Well, I guess my job here is just to smile and nod*—which I did.

Patience and friendliness mean you have to suffer some fools; you have to smile and nod and wait until the joke exhausts itself or the hour is up. You have to listen to the whole sales pitch and maybe by the end even buy something out of politeness. Men and women both do these things, to be sure. I know that Professor "Lincoln" deployed his schtick with some male interviewees, too, but the claim is that it is more important for women to respond graciously than it is for men to respond that way. The expectation that women will humor difficult people or agree with nonsense, especially from men, is descriptive and prescriptive. Women are regarded as odd, if not outright hostile, if they don't play along.

It's not controversial to say that women are more vigilantly policed for their warmth. Passersby admonish women who are otherwise thinking their own thoughts to "smile, honey," and there is of course a whole acronym for women whose neutral facial expression does not project friendliness (the "Resting Bitch Face," or RBF). Even the word "bitch" is a specific indictment of a woman's failure to be a good sport. There is an expression that I have always understood as a compliment, that someone "does not suffer fools gladly." People who are supposed to be friendly and patient (ladies!) are almost definitely going to have to suffer some fools and appear pretty glad about it.

If we move the analysis up a level, this same dynamic has implications for more consequential choices around social cooperation. Caring for others has historically been women's work, though in some respects that division of labor has changed over the last half century. Legally, women are no longer excluded from economic participation or consigned to the family sphere. Access to the labor market, birth control, no-fault divorce, and child support laws have dramatically reduced the formal constraints on women's choices. (Many formal legal barriers remain, and reproductive freedom has recently been dramatically curtailed, as I discuss in chapter 8.) But there is still an underlying sense that women, not men, are going to have to be the ones to take the raw deal when the social safety net fails.

Early in my career, I sat in the audience of a faculty workshop on insurance law. The presenter started a discussion on the costs of long-term family care: What are the financial, professional, and social costs of caring for adult family members, siblings, or aging parents, who become disabled or ill? The conversation naturally turned to the gender realities of care work: descriptively, women still spend more hours a week, over months and years, invested in long-term caregiving, even outside of parenting. One of the men in the audience raised his hand and asked why we need to worry about people—women—who commit to care that they are not legally obligated to assume? If women *choose* to do care work and men *choose* not to, why would we want an insurance product that mitigates the consequences of a *choice*?

I was a junior faculty member at the time and kept my counsel, but I was also the mother of two young children, so I did know something about how family care generally is perceived in our society—and it's really not perceived as optional for women. (If you want to test this in a little thought experiment, imagine a situation in which a parent of school-aged kids takes a one-year job overseas, visiting home for a week every other month. How does our society judge a father who makes that choice? A mother?)

Men and women of equal economic stability—already a stretch, but let's go with it for now—will pay different *social* costs for failing to assume unpaid family labor. If your kid goes to school in weird clothes or you forget to pack lunch, that reflects poorly on the mother in a way that does not seem to even touch the father. There is an entire feminist discourse on how wives are expected to care for disabled husbands for years, while husbands are heralded as heroes if they manage to last a week of spousal caregiving without adultery and abandonment. Women who don't tend personally to care work are selfish; men are sensible.

In my formative late adolescent years, there was a recurring image on the nightly news of a woman, usually a wife, standing silently beside her husband as he publicly apologized for being unfaithful. Hillary Clinton, Silda Spitzer, Elizabeth Edwards—these women of a particular generation offered public forgiveness for personal betrayals. They stood by their cheating men. That extraordinarily personal decision is fascinating and complicated; it requires one member of an intimate relationship to agree to play the public sucker. What's more is that the women were socially rewarded for their choices. Hillary Clinton won over skeptical Americans when she submitted to public humiliation; playing the fool was her ticket to (partial, fragile) likability.

Clinton's advisors correctly predicted that women who accept the sucker's payout get access to a variety of social and economic rewards: a lasting marriage, a book deal, good polling. This itself is a form of sexism, the isn't-she-a-saint condescension, that reinforces the narrative of the female dupe.

The social psychologists Peter Glick and Susan Fiske describe this attitude as "benevolent sexism," the idea that some gender prejudice comes in sheep's clothing. It looks complimentary, but it has a pernicious edge. They designed a questionnaire to measure benevolent sexism and to test their prediction that even superficially admiring statements could be correlated with discriminatory behaviors. The items on the scale included things like "Women, compared to men, tend to have a superior moral sensibility" and "Women should be cherished and protected by men." It was a logical outgrowth of sex stereotyping from Bem forward: women (tender, caring) deserve male protection—and probably require it, too (childlike, naïve).

But Glick and Fiske went further. Sexist attitudes about women, they pointed out, are not just, or even primarily, about chivalry and pedestals. Sexism is about patriarchy; it justifies active economic subjugation and physical domination. Glick and Fiske added a set of items to measure "hostile sexism," and those statements were notably darker: "Once a woman gets a man to protect her, she usually tries to put him on a tight leash" and "Women seek to gain power by getting control over men." Specifically, a number of the statements on the hostile-sexism scale are about women who trick men to gain status. They include:

❖ There are many women who get a kick out of teasing men by seeming sexually available and then refusing male advances.

❖ Women exaggerate problems they have at work.

❖ When women lose to men in a fair competition, they typically complain about being discriminated against.

❖ Many women are actually seeking special favors, such as
hiring policies that favor them over men, under the guise
of asking for "equality."

❖ Feminists are seeking for women to have more power
than men.

These statements describe specific sucker's ploys: telling a man
you are sexually available and then rejecting him; telling people
you have problems at work to get sympathy when you have no
problems at work; claiming discrimination when there is none; and
claiming you want equality when you really want "special favors"
and power.

Stereotypes about women's place in society are deeply rooted in
a scammer-dupe framework. Hostile and benevolent sexism seem
to be measuring two different sets of stereotypic beliefs: women
are pure and moral *or* women are liars and cheats. In fact, though,
they are beliefs that go hand-in-hand, culturally and statistically.
Benevolent sexism describes the rewards of playing the sucker,
and hostile sexism threatens the costs of refusal. The legitimizing
myth of the beautiful little fool promises that women are getting a
good deal—men "cherish" them and "put them on a pedestal"—
and preemptively marginalizes the complainers as liars and teases.
Just as we saw in the context of race, hostile sexism espouses the
belief that women who make claims for equal treatment are actually trying to jump the line.

Her Cheating Heart

You don't need to be a social psychologist to know that misogyny
is always ready to pounce at the prospect of a duplicitous woman.
Being cheated by a woman is deeply offensive to traditional male
dominance, because it means being fooled by a fool; the severity of
that status threat means that women are fiercely surveilled.

The extraordinary vigilance against duplicitous women has roots, symbolically and possibly biologically, in heterosexual reproduction. For evolutionary psychologists, the reason that men vigilantly police women's sexuality is because the vigilant men are the ones who accurately invest resources in their *own* genetic offspring. A mother knows her baby reproduces her family's DNA; a father has to take the mother's word for it on a baby-to-baby basis, or so goes the theory. In terms of evolutionary strategy, his vigilance about his sexual partner's fidelity is naturally connected to his reproductive success. Part of winning at evolution is to have the fear of playing the sucker turned to high alert.

Whether or not natural selection has actually favored male sugrophobia, the idea is embedded in narratives of masculine sex roles. What men cannot be is weak, gullible, and impressionable. You don't lose status points by being overbearing or domineering, but the prohibition on being fooled is fundamental to the nature of masculine prescriptive stereotyping.

Kate Manne, the philosopher and author of *Down Girl*, has argued that the fear of deception is the key to the logic of misogyny. Women's social roles require them to support men, with love, loyalty, caregiving, sex, or children. She points out that "because of women's service position, their subordination often has a masked quality about it: it is supposed to look amicable and seamless, rather than coerced. Service with a smile, not a grimace, is the watchword." But, argues Manne, the desired service—giving—puts men on constant guard:

> He will also need to be assured of (a) her honesty, (b) her loyalty, and (c) her constancy to be assured of such goods as security, stability, and ongoing safe haven . . . It's not a safe space for him, if she might already have one foot out the door, or loves him only conditionally on his worldly success, his good reputation, fame, or similar.

Women have to play the fool—adoring and sanguine—but men have to remain on guard, because they can't *really* ever know if they are getting the genuine respect they want. The story has the same structure as the evolutionary account: the very goods that are demanded are the most difficult to verify. The man is stuck policing authenticity even when he is by all accounts getting what he wants.

When I teach the law school's course on wills and trusts, I find that most of the cases in the textbook are family disputes. The people who sue over property passed at death are usually the intimates of the decedent, because those are the people who feel jilted when they don't get what they expected from the estate. One of the most predictable triggers for contesting a will seems to be the existence of a stepmother. Family members go to court because they are angry that their relative—their father or grandfather (most of the people leaving behind real money are men)—was fooled into giving money to an interloping woman who claimed to love him. It is a repeated pattern: The husband and father leaves his estate to his second wife, a younger woman who is not the mother of his adult children. The adult children sue, arguing that their father was unduly influenced, or incompetent, or poorly advised. Had he seen the situation clearly, they claim, he would have left his money to them. No matter the specific claim, the subtext is always the same: she was a gold digger. He was taken in by her offers of access to her body, her flattery, even her caretaking. He never would have left her this estate if he had understood that he was playing the fool.

I generally try to reject this sexist framing altogether, but even if I were to be convinced to pigeonhole women as undeserving opportunists, I'd have to really want to see duplicity for these cases to fill that bill. Students often have in mind a prototype of Anna Nicole Smith and J. Howard Marshall—you may recall the media frenzy surrounding this one-year marriage between a young woman and an eighty-nine-year-old millionaire—but real cases are much more complex (and even that case is way more complex than its popular gloss). I occasionally feel compelled to stop class discussion and reiterate the facts of the standard cases: This couple was married for thirty years! The decedent was ninety-five! The "gold-digging schemer" in question is an elderly woman who has been a caregiver for years. After all that, what would it even mean to untangle devotion from manipulation, authentic love from artifice?

Sex

For a woman, much more so than a man, the mere fact of a sexual relationship raises the possibility that she is a dupe, a conniver, or both. Paradigmatically, if heterosexual sex is happening, a woman is "getting screwed." The target of seduction, the lady, is a presumptive sucker. What a fool; she can't say no; she gave the milk away for free. (The word "sucker" in the sense of "dupe" did not originally refer to a sexual act—apparently its etymology has to do with a suckerfish that was very easy to catch—but for modern readers there is obviously a sexual overtone.) But women who do not offer their services with a smile come under suspicion, too: a woman who offers sex but really wants money is a gold digger; a woman who says yes and then no is a cocktease; a sexually active woman who pretends chastity or monogamy is something even worse.

Even when women have not consented to sex, we are all familiar with the tropes that insist they are marks rather than victims per se. The victim of sexual assault who flirted, drank alcohol, wore the wrong clothes, or walked down the wrong street may still be dismissed as a party girl who "asked for it." Although some norms have evolved, we still see a persistent impulse to describe sexual coercion in terms of the victim's choices. She may not have consented to sex, but she consented to meeting him at a bar, or drinking, or going back to his place. In their own way, these narratives are the legitimizing myths of patriarchy, too.

There is a natural sucker narrative to sexual assault, and it is usually the woman starring in the role of fool. But lurking in the background of every campus sexual assault scandal and long-form think piece about "Has #MeToo Gone Too Far?" is something else: the documented, widespread fear that increased institutional protections for women claiming sexual assault will make men vulnerable to lying schemers. Fears of "he-said-she-said" disputes often describe women luring men into sexual complacence, engaging in consensual activity, and then fraudulently claiming harm (note the derogatory "crying rape" often used here and meant, presumably, to evoke "crying wolf") for the purpose of targeting their mark.

In the last ten years, two cultural phenomena spooled out together. There were widespread, partly successful calls for increased accountability for men accused of sexual harassment and violence, from #MeToo to legal reforms to campus disciplinary procedures to the Shitty Media Men list. At the same time, some high-profile sexual assault allegations—Duke lacrosse, a discredited *Rolling Stone* story about a UVA fraternity—proved to be false or unsubstantiated, to enormous public dismay. These converging narratives fed a set of popular fears that women were being handed too much power to hurt men by lying about sex.

Take the following as an example: the *Atlantic* ran a three-part story with a heart-wrenching, infuriating tale of a mistreated male student accused of sexual assault with scant evidence of wrongdoing and no recourse to defend himself. People were outraged. University faculties wrote open letters demanding fair process for accused students. Kate Manne coined the term "himpathy" to describe the groundswell of moral feelings of sympathy toward alleged male perpetrators of sexual violence. Many opponents of campus sexual violence reforms had these really vivid fears that men would be made miserable marks—fooled and then expelled, canceled, unemployable. That scam—women lying to bring down trusting men—is disruptive to the normal order, which is perhaps why it generates so much retaliatory aggression.

Even the repeated use of the language of "he-said-she-said" felt like a dig, the jaunty syntax too close to his 'n' hers robes or a Jack-and-Jill bathroom. On its face, "he-said-she-said" just means that there are two incompatible stories and both are difficult to verify with extrinsic evidence—but the implied equivalence can feel disingenuous. Most of us realize that when "he" and "she" make competing claims in this world, he's the one who gets deference. And the language gestures, however obliquely, to an assumption that women have standing reasons to lie about sex, wanted or unwanted.

One way to neutralize women who make claims of sexual assault is to turn the narrative of duping around on them. While the #MeToo movement pushed forward, the journalist Elizabeth Bruenig looked back at an incident from her own Texas hometown. A sophomore cheerleader accused three football players of a brutal rape; she was received with skepticism and, ultimately, revulsion. The girl—the attested and plausible victim of a violent sexual assault—brought a terrible, disruptive claim: that popular high school boys had committed a serious crime and lied about it. She told a consistent story, had physical evidence of force, was clearly traumatized, and yet, "the tone of murmurs around the school indicated that students believed the exact opposite: that [the girl], perhaps intoxicated, had agreed to sex and then regretted it, and that, in accusing the boys of rape, caused trouble not only for herself but also for her classmates . . ."

She was ostracized; she dropped out and then got her diploma from an alternative school. The slip, from prey to predator, can be so quick, it's almost sleight of hand. She was not victim but perp, not mark but scammer.

For me, the article shed a retrospective light on a confusing episode from my own adolescence, a story of sexual predation that ended in the arrest of our high school health teacher and the bewildering departure of the girl he had tried to abduct. The story unraveled in real time for us, of course, but it also caught threads from other encounters and earlier years that we had not quite observed as a pattern.

My freshman health teacher, a forty-eight-year-old man with buzzed gray hair and an enormous paunch, had almost every ninth-grade student in his mandatory class. He was also the driver's ed instructor, a fact so trite in this context that it's practically a meme.

He was responsible for sex education, along with an idiosyncratic health curriculum I assume was of his own design. Some of his teachings were notably dark. He taught us to beware of a peculiar female pathology of being "accident-prone," something that he described as happening to women who want to cause their own pain and get attention for it. For some reason there was a whole lecture one day on laxatives as an ineffective and inadvisable form of eating disorder. "You really don't even lose weight that way," he said sternly to a room full of people who could only go to the grocery store if their parents drove them and had until that moment never considered purchasing, much less abusing, over-the-counter laxatives. He was married to his fourth wife at the time. I think we were dimly aware there were accusations in his past. His name, which I am not making up, was Mr. Dick.

He liked to remark on our outfits in class. "You're looking 'foxy' today," he might say to me if I wore a skirt—casually, as though we shared some sort of wry understanding. We did not; I was thirteen.

Mr. Dick was genuinely beloved by some, an overlooker of bathroom cigarettes, a fudger of attendance records. He was proud of his social capital and boasted that girls would come to him—we could come to him!—if they were worried that they might be pregnant. He would bring out a calendar, show them how to count the days. When did they last have sex? And when was their last period? (Not dissimilar from the questions at obstetric intake.) He could reassure these girls, he told us, although it occurs to me now that I don't know what he counseled if the math was not reassuring. Teenage parenthood was common enough that the high school had a day care on the premises for the children of students. We called it the Baby Barn.

One afternoon in my senior year, some of us came to understand that a friend, a younger girl I'll call Jen, had cut her classes, and Mr. Dick was also nowhere to be found.* It's hard to say how the information traveled and congealed, but it did. We didn't have cell phones, but we knew very quickly what was happening and just how bad it might be. Jen's best friend was a girl named Angie, a funny, scrappy junior whom I had befriended on the Envirothon team. She came to my house after school, where we pondered the situation. At one point Angie actually looked up Mr. Dick's number in our phone book and called him. His wife answered. No, Mrs. Dick did not know where her husband was and had not seen Jen. No, she had not heard anything else.

Jen was found before the day was out. We heard that the police had pulled Mr. Dick's car over less than an hour from the high school. She and her mother talked to the police and cooperated with the prosecution. A year after the attempted abduction, I read a news clipping in which a reporter interviewed Mr. Dick, who was awaiting sentencing. On release, he was hoping to become a long-haul trucker.

Jen came back to school eventually, with a haunted look to her. Older students passing her in the hall would hiss epithets when she walked by. She soon transferred to what we understood to be some sort of special school—and it was that parallel detail that snagged for me when I read about the Texas case. At the time, I heard a grown adult—a teacher or a parent; I can't remember—mutter that Mr. Dick was a fool who let a girl "lead him on."

As teenagers, this verdict was somehow both predictable and surprising at once. We knew that girls don't come out of these things unscathed, even as we knew that this was the rare case where a man would face consequences, too. With her abductor in custody, Jen should have been reclaiming her place in her social world. But between the time she got out of his car and the time she walked back through the school doors, a local narrative had flipped: she had tricked him. The mark became the operator.

---

* Names of minors and some details have been changed for privacy.

❖

The fool's warning is constricting for all of us. Its requirements for men are severe and exhausting: *En garde*, or else! For women, the looming threat of inadvertent participation in a scam, on either side of the deal, is more complex but no less fraught. Cultural attitudes toward female fools are deeply ambivalent. On the one hand, the flatterable, the easily misled, the targets of the "baited come-on"—these women get our pity, not our admiration. On the other hand, some of that foolish naïveté is bound up in heteronormative ideals, or at least romantic performance, where getting screwed is sort of the point. Less glibly, people whose social acceptability depends on them being friendly, patient, and compassionate will sometimes get boxed into a sucker's corner because the social costs of refusal are too high.

When women refuse to play along, they open themselves up to the converse accusation: if you won't play the game, maybe you're running one. It's a set of constraints that leaves very little room for error; it puts women on a tightrope.

When my young son was mysteriously ill and we were scrambling for a diagnosis, I remember the chill of suspicion as I tried to convince some long-suffering specialist that my miserable child was, like, *clinically* miserable. Either my kid was having me on or I was having the doctors on, but there wasn't much other choice for me, just sucker or con.

With these dual accusations in mind—she's a fool; she's a fraud—I recently revisited an article by the psychologists Pauline Clance and Suzanne Imes. The article, written while they were graduate students, was called "The Impostor Phenomenon in High Achieving Women." It may not be a common household read, but most of us know its popular shorthand, the impostor syndrome.

The impostor syndrome describes the feeling, largely identified with women, of being fraudulently successful. The authors argued that the impostor phenomenon—their preferred term, since it is not a diagnostic category—comes from the internalized sense of self as inadequate: "Given the lower expectancies women have for their own (and other women's) performances, they have apparently internalized into a self-stereotype the societal sex-role stereotype that they are not considered competent." Almost a half century later, this interpretation has stuck, with not a little irony: women are so easily misled by sexist stereotypes that they don't know it when they're actually smart!

Maybe what women are picking up in their impostor anxiety is not just the cultural narrative of incompetence but also the cultural narrative of duplicity. It is one of the stereotypes embedded in misogyny: that women will exaggerate and lie to get special favors; that they will pretend to be nice to get money; that they will pretend to be honest to get love. To confuse the issue even further, women get social rewards for pretending *not* to be talented and successful, fending off critiques of arrogance or entitlement.

A successful attorney I know graduated from law school in the early 1970s, and she told me the story of her professor pointing at the few women in the room and announcing that they should each be trying to justify, every day, that they had taken a man's seat. The impostor phenomenon is a fear from within (Am I a fraud?) entangled in a threat from above (Ladies, you're probably frauds). Making successful women feel like impostors is the logical endgame of sexist sugrophobia: if you don't feel like a fool, maybe you should feel like a fake.

CHAPTER 7

The Cool-Out

On November 2, 2021, hundreds of Americans gathered on the streets of Dallas, Texas, to cheer the political return of John F. Kennedy Jr.—the same JFK Jr. who was in a deadly plane crash off of Martha's Vineyard in 1999, a tragedy that killed him, his wife, and his sister-in-law. As followers of the QAnon conspiracy, the gathered fans believed that he was going to reveal the truth of his faked death, reinstate Donald Trump to the presidency, and accept the offer to serve as vice president himself.

None of this came to pass. Indeed, QAnon has made a number of high-profile predictions that have failed. Comet Ping Pong in Washington, D.C., proved to be a family-friendly Italian restaurant and not a site of child trafficking. Joe Biden's inauguration in 2020 was not disrupted by mass arrests and executions. Hillary Clinton was not "extradited" or arrested on any of the many predicted dates in 2017, 2018, or 2019.

A reporter from the *Dallas Morning News* interviewed one of the Kennedy-Trump supporters as she waited for the announcement that wouldn't come, asking her how she would react if the younger Kennedy failed to appear. Although she was ostensibly expecting to see a man who had died decades earlier, in some respects she was quite self-aware. "We'll figure that something happened in the plan that made it not safe to do it," she said. "If it doesn't go down how I believe it will, that's OK. We'll figure it just wasn't the right time." She could forecast her own resistance to changing her mind; she was not going to quit QAnon, or demand answers, or change her behavior at all. No amount of substantiation or logical reasoning or evidence that the Kennedy story was a hoax was going to convince her that she had been duped.

The perverse logic of sugrophobia is that being duped is so aversive that anything seems better than admitting to being taken in. Indeed, the specter of the sucker can loom so large that those most afraid of being fooled are the ones least able to name it when they see it. It's like a late-night infomercial purchase that you'll later insist was a great deal, even after it's clear you've been had. If you think it's really shameful to be a dupe, you don't want to admit when you get ensnared—and everyone gets ensnared sometimes.

The way we interpret sucker dynamics in real life is malleable and shifting; a transaction can seem innocent or malevolent, depending on the context or the players; this is why the concept is so vulnerable to weaponization. That same malleability opens up another possibility: that a sucker story might be *un*told to placate victims of exploitation. You might feel betrayed, but it's just business. You weren't duped; you just misunderstood what I said.

Erving Goffman, our favorite mid-century sociologist of suckers, was obsessed with the ways that people could talk themselves out of awkwardness. He saw the con as the ultimate test. It is *so* embarrassing to be suckered; it's a huge psychological challenge to recover. What social and mental gymnastics are employed? Recall his setup: "In cases of criminal fraud, victims find they must suddenly adapt themselves to the loss of sources of security and status which they had taken for granted." Goffman's essay on "cooling the mark out" dealt with these two key facts: the harm is serious—

what could be a bigger deal than the loss of security and status?—
and it is followed by an urgent imperative to adapt. But how?

Cooling the mark out is redefining the situation "in a way that
makes it easy for him to accept the inevitable and quietly go
home." It's a reorganization of the sense of self that needs to happen quickly or else it will threaten the scheme:

> The mark is expected to go on his way, a little wiser and a little
> poorer.
>
> Sometimes, however, a mark is not quite prepared to accept
> his loss as a gain in experience and say and do nothing about
> his venture. He may feel moved to complain to the police or
> to chase after the operators. In the terminology of the trade,
> the mark may squawk, beef, or come through. From the opera-
> tors' point of view, this kind of behavior is bad for business . . .
> In order to avoid this adverse publicity, an additional phase is
> sometimes added at the end of the play. It is called cooling the
> mark out . . . The operator stays behind his team-mates in the
> capacity of what might be called a cooler and exercises upon
> the mark the art of consolation.

Goffman saw the canonical scam—the mobster who targets
foolish dupes looking to turn a quick buck—as a prototype, and
also a metaphor. It's about the chump in a street hustle and it's also
about the exploited worker in a breach of the social contract. The
art of consolation can be practiced by the customer service repre-
sentative at Wayfair or a chagrined hiring manager who calls with
bad news. It might be a soothing excuse ("You're just really over-
qualified") or a conciliatory sweetener ("We'll waive the shipping
charges for your trouble"). But Goffman's insights apply more
deeply if we think of cooling as a process that usually unfolds in-
ternally, in response to the social world.

Deeply suspicious people—that's most everyone—nonetheless
permit a wide range of hustles, scams, and injustices to be perpetrated against us. Indeed, rationalization is predictable: there is a

pattern to the stories we tell ourselves and each other to make some of the most intractable raw deals psychologically manageable.

When Goffman wrote about the job of the cooler, he described an agent who could do the work of talking someone off the ledge. But we might think of this agent as the last leg of a relay, finishing off a job that has already been started. He brings pressure to bear in the form of threats, promises, or little payouts—situational tweaks to make the problem go away. The reason his tactics work, though, is that the human mind is its own most effective cooler.

Peer Pressure

There is a reason that breakups often happen in busy restaurants: sometimes the coolers are just other people, and all they have to do is be there.

To restate the obvious, humans are exquisitely attuned to social cues, and norms, and expectations. We might ask why we're vulnerable to cooling in the first place, and one answer is that we actually prefer the result: the choice to accept the inevitable is easier and more natural, socially, than making a fuss.

I understand this urge on a deeply personal level, because fear of confrontation is one of my own most prominent and most frustrating personality traits. I was at one point in my career a deputy dean, which is a bit like being a vice principal in a high school, only the job rotates every two years, so it's an assignment rather than a career. It's not a role people usually like to have, and in fact some colleagues helpfully reminded me that I was a sucker to have agreed to it at all. During my stint I was embroiled in an extraordinarily contentious faculty dispute. In addition to regular hurt feelings, we briefly became a media focal point for competing narratives of academic freedom and campus speech. One faculty member, accused of insensitive comments, responded with claims that the controversial protected speech was being "silenced" by a variety of institutions and people—including me. (For the record, that was not my intent.)

The aggrieved professor convened an all-school town hall of sorts, inviting students to a lecture and Q & A about the controversy, which by then had made national news. In my administrative role, I helped arrange security—there was concern that protestors from outside the law school might try to shut it down—and helped set up the logistics for a large group event. Then I grimly took my place in the audience.

About three-quarters of the way through the lecture, my unhappy colleague announced from the podium that there was a special person who deserved the title of "anti–role model" in this whole mess, and then read my name out. Actually, there were two of us cited for this dishonorable mention, but, like most faculty, the other person had wisely steered clear of the whole event. And what *I* did while a colleague told a full auditorium, including almost all of my own students, that I was *not their role model* was . . . nothing. I sat paralyzed while the lecture continued through the hour.

I spent half the time trying to figure out if it would be rude to raise my hand to interrupt the talk or if instead I should queue at the microphone. Would I be violating the norms of my administrative position if I spoke publicly? Would the students feel I had usurped their time to ask questions? It had a very "Do I dare to eat a peach?" sense of dithering; while I parsed the norms and niceties, the event barreled on and then ended, and I did, ultimately, quietly go home. Like the rejected partner in the busy restaurant, I would not, and did not, make a scene in front of three hundred students.

Although the tendency varies from person to person, one basic human instinct is the preference—all else being equal—to go along. Sometimes we call it obedience and sometimes we call it conformity, but it is fundamentally the same thing. This is what the psychologist Stanley Milgram was researching when he invited people to shock each other in the basement of a Yale laboratory in 1961, and this was what psychologist Irving Janis was talking about when he coined the term "groupthink" after the Bay of Pigs disaster: the unthinking motivation not to rock the boat.

The research giants of this chapter—Milgram, Janis, Goffman,

as well as Leon Festinger, Solomon Asch, and Melvin Lerner—
were doing groundbreaking research during the 1950s and 1960s,
a time when academic psychologists especially were grappling
with how to think about human psychology in light of the Holocaust. These were not just researchers studying evil acts per se;
instead, they were studying when regular people will accept injus-
tice. When will they go along with group violence that they would
not otherwise have chosen? Why will they obey commands they
know to be morally corrupt? Their work represented a paradigm
shift in psychology—a shift from asking how individuals affect
their social worlds to asking how social worlds affect individuals.

Solomon Asch, a psychologist from the Gestalt tradition,
brought to social psychology the Gestalt precept that the whole
is greater than the sum of its parts. This was his view of behavior,
too, that "social acts have to be understood in their setting, and
lose meaning if isolated." Asch was interested in social influence,
and in his most famous experiment he wanted to demonstrate the
unequivocal effects of social pressure on people who suspect they
are being played for fools. He set his subjects up with a simple task.
They saw two cards. Card 1 had a single line, and card 2 had three
labeled lines. The task was to match the line from card 1 with the
corresponding line from card 2.

Participants were brought in to perform this task in groups of
eight or nine men at a time—the study was performed with Harvard undergraduates in 1955, i.e., all male—but in each group,
only a single person was actually an experimental subject. The rest
were "confederates" in that they were in on the game, following
predetermined instructions. Asch or his research assistant would
show the cards and ask each subject to reply in turn to the question
"Which line is a match?" The first few rounds were normal. The
task was very easy and everyone just announced their answers. On
the third round, though, the first six people seemed to choose the
wrong line, as they had been secretly instructed. Now the question was what the actual subject, placed at the end of the response
queue, would do.

It's helpful to put yourself in the shoes of the guy at the end of the row. He is looking at three lines that are different lengths. He has a reference line, which is clearly the same length as line A. This experiment has been uneventful, and the task is easy. But now six people in a row have just announced that the reference line is a match for line B, a line that is in fact much shorter. What would you do? What would you expect others to do?

Some people just kept giving their true perceptions, accurately selecting the congruent line as before. Over a third, though, conformed to the "misleading majority." Asch's reflections on his conforming subjects identified some who went along basically to be polite and others for whom the whole interaction caused real self-doubt:

> [Some] yielded in order "not to spoil your results." Many of the individuals who went along suspected that the majority were "sheep" following the first responder, or that the majority were victims of an optical illusion; nevertheless, these suspicions failed to free them at the moment of decision.

The first pattern Asch saw was that some of his subjects were going along with their classmates because the social setting made it unpleasant to buck the trend. Choosing the correct line would have seemed to be accusatory: The rest of you all are defective, or lying! And just as Goffman, Milgram, and many others predicted, a smoothly functioning social setting was an incredibly effective enforcer of conformity. The scam was plain, laid out with preschool-grade clarity; the consequences for rebellion were low; but still, many of the marks just went along, without a squawk.

Once a social interaction is underway, an accusation is disruptive, awkward, and embarrassing; sometimes it's worth taking one on the chin rather than creating discord. In fact, an ingenious study recently updated the work of Goffman and Asch, homing in on the specific—and specifically aversive—disruption of calling out a scam.

Sunita Sah, Daylian M. Cain, and George Loewenstein are professors who study, among other things, the pressure to take bad advice. In particular, they have focused on what happens when people get advice with ulterior motives. This might seem like an esoteric dilemma, but actually it happens all the time. Oftentimes we need guidance or consultation, but the experts we'd otherwise turn to have financial incentives to give self-interested advice. (Think of Audrey Vokes getting feedback on her dancing from people who wanted her to keep giving them money.) Does that tooth really need to be sealed? Do you need new brake pads? Does the chimney *really* need to be relined? How on earth would you know? The mechanic who knows how a transmission works also does better if I buy more services. The same is true for the dentist and the HVAC company, not to mention financial and real estate brokers whose incentives are just slightly misaligned with their clients. To be tendentious, this is also true of other structures of capitalism: What is advertising, for example, but bad advice from a biased source?

The way that we regulate these conflicts of interest is usually to require them to be disclosed. If my doctor knows I need a hip replacement and recommends a particular prosthetic, she might be legally obligated to say something like "I should tell you that I have a stake in this company, so I directly profit from sales," if that is the case. Perhaps she adds, earnestly, that she would never let that affect her advice, but she wants me to have the full information so I can make my choice.

Sah, Cain, and Loewenstein wanted to know: Does it really work to tell people that if they are worried about being scammed, they can just . . . opt out? Their first hypothesis was that the disclosures would be ineffective, and people would not opt out. They did find that, sort of, but it was even more perverse. When people were worried about being misled, they were *more likely* to go along with their scammer.

If we think consumers are naturally afraid to be taken for a ride, it seems logical that when the doctor says, "I'm recommending a product that profits me personally," the patient will object, or walk out, or at least ask for a second opinion. The doctor has just disclosed that she's possibly running a scam! But most of us know that we wouldn't react like this. Policing counterparties for conniving behavior is a big social deal; it conveys not only distance but disrespect. Especially if the advisor is someone who feels high-status—a doctor, a financial advisor, a white man in a suit—an accusation, even one that's indirect, is very insulting to them and very uncomfortable for you. There is strong social pressure not to accuse other people of being scammers.

Sah, Cain, and Loewenstein replicated the advice-disclosure-choice interaction in a lower-stakes but more research-friendly context than prosthetics and car repairs, by using a little lottery game. They had a trained research assistant hang out on the ferry from New London, Connecticut, to Long Island, New York, and ask passengers to fill out a short questionnaire about the ferry service. The questionnaire itself was a decoy: the experiment didn't come until the subjects got their payments for filling it out.

Each of the subjects was told that they could get $5 for taking the short survey; or, if they preferred, they could enter into a lottery in which, depending on a roll of the dice, they could earn as much as $10 or as little as $1.

The lottery was objectively not as good as the $5 sure thing: the average payment was set to hover just below $5. Subjects in the control group were offered the lottery with no other information, and under that "no-advice" condition they only chose it 8 percent of the time; people overwhelmingly favored the $5 payment. But for other passengers the research assistant would offer some advice, depending on their randomly assigned experimental condition. In the "conflict" condition, he would point to the lottery and say, "I've seen a bunch of the payouts of the drawing, and I suggest you go for that option; it often pays nicely." For passengers who heard only that line, about 20 percent chose the lottery—a small but significant increase.

A third condition, "disclosure," got advice *and* a disclosure.

Before launching into his suggestion to choose the lottery, the research assistant started with "I should tell you that I get a small bonus if you pick the drawing. That said . . ." Passengers who heard the advice *plus* the disclosure chose the lottery 42 percent of the time. Why would it be that people have a harder time turning down bad advice when they have been affirmatively warned off?

The researchers called the effect "insinuation anxiety." People feel awkward or uncomfortable if they think that declining advice will be interpreted as an implicit accusation of bias or corruption. After the lottery choices had been made, the research assistant asked subjects to complete a post-questionnaire study, to drill down into what caused the perverse effect of the disclosure. Subjects who took the conflicted advice were more likely than others to agree with the statement "I was concerned that the interviewer would believe that I thought he was biased if I turned down his recommendation."

Once a transaction like this has started, there is no way to get out of the interaction without either agreeing or accusing. People who weigh the social costs of accusing higher than the personal costs of agreeing will indeed cool themselves out; in the language of Goffman, they "accept the inevitable and quietly go home."

In this experiment, by the way, the perverse effect was strongest for women. The authors noted that the research assistant was a middle-aged man in a business suit.

The Pressure Is Coming from Inside the House

When Solomon Asch observed his peer-pressured line judgers, he noticed that some of the conformers were doing more than just going along to get along; they were changing their internal beliefs to match the situational cues. He found this reaction troubling:

> More disquieting were the reactions of subjects who construed
> their difference from the majority as a sign of some general de-

ficiency in themselves, which at all costs they must hide. On this basis they desperately tried to merge with the majority, not realizing the longer-range consequences to themselves. All the yielding subjects underestimated the frequency with which they conformed.

Social pressure can convince the mark to accept his fate with resignation, knowing that the fight is not worth the candle. Other times, though, the process of cooling out is more cognitive than behavioral, a true transformation of self-persuasion.

It's not hard to imagine how this happens. A couple of years ago, a group of students in my consumer law class did an excellent presentation on consumer privacy. Half of their slide deck was devoted to just the information being collected by one dictionary app for the iPhone. The default settings permitted it to track and record your location and, like, befriend all of your contacts. It was horrifying. But when I think of a typical day on the internet, I just wander through that thicket without a second thought. When I'm doing a lot of research, for example, I come across a button that says something like "Accept all cookies" or "I agree to the Privacy Policy" multiple times a day, sometimes literally every ten minutes. If I reflect on it, I have signed away almost all of my data privacy rights over the years.

If you asked me to think very seriously about my values and preferences, specifically with respect to data privacy, I would tell you that data privacy is important to me. Unfortunately, it is also true that I keep agreeing to sell it for basically nothing. One interpretation of my situation is that I am a sucker. I just keep assenting to terrible privacy deals while companies get rich off of my carelessness.

I don't want to feel like a sucker, though. I value data privacy; I have signed away my rights to data privacy: these two dissonant statements create psychological pressure. That sets up a new situation where I am looking for a way to relieve the cognitive dissonance. The reason the data privacy beliefs feel bad is because the implication is that I am a pawn, giving up something I value for scant rewards. So what can I do? I can become a data privacy fanatic, quit Facebook and Twitter, refuse cookies, eschew Google. This is unrealistic, probably, because it would be onerous and limiting. I can live in rage and disappointment, of course, but that's just dwelling in the pressure rather than relieving it.

If I can't do much about my data privacy—I submit to you that few of us can—and I don't want to reckon head-on with my fundamental inability to control unseen pernicious forces, there is still one option left for me: on second thought, maybe I don't care about data privacy as much as I thought! I can't change cookies but I can change my beliefs. This is the psychology of cognitive dissonance, which describes the internal pressure to resolve incompatible cognitions.

The roots of cognitive dissonance research are in dupes. The social psychologist Leon Festinger originated and developed the theory of cognitive dissonance over decades of experimental research, but his early interest developed while he observed members of a doomsday cult.

Festinger and two of his colleagues from the University of Minnesota, Stanley Schachter and Henry W. Riecken, had heard of a woman in Illinois who claimed to be in contact with aliens. The aliens, known as the Guardians, announced through Dorothy Martin of Oak Park, that they would be arriving in December and would be rescuing their followers while obliterating everyone else with a massive flood. The believers gathered outside of Martin's house on Christmas Eve and waited for the flying saucer that was due to pick them up at 4:00 p.m. Festinger, Schachter, and Riecken wanted to know what would happen when the Guardians failed to arrive.

The researchers conceived of the dilemma as a kind of equation. What happens when the believers have two simultaneous conflicting cognitions: (1) I thought the world was going to end by 4:00; and (2) it's time for dinner and we're all still in Dorothy's driveway. What third statement reconciles those two facts?

One possibility is to solve the equation with (3) I'm a chump;

my belief was wrong. Generally, that is not a popular solution. As Festinger observed, most people will opt for (3) the aliens got the time wrong and they'll be here later; or (3) I never said I believed it was definitely 4:00; or (3) I think Edna over here violated one of the rules and now the aliens are not going to rescue us as planned. (There was some discussion, apparently, about whether or not the aliens' prohibition on metal included bra straps and silver fillings.)

Festinger and his colleagues, notably the psychologist Elliot Aronson, observed a pattern: people are suggestible, but they don't like to admit that about themselves. Like the Q supporter waiting for JFK Jr., people will jump through mental hoops to avoid the conclusion that they've been duped.

In Festinger's early experiments, he put his subjects through ordeals to induce dissonance. He posited that when we experience a harm that seems to be the result of exploitation or betrayal, we will fill in the equation with anything else besides "I'm a fool"— including, most saliently, "I didn't experience a harm at all; I liked it." When people go through a hazing ritual, for example, they come out announcing not "Wow, I was pressured into doing some really humiliating stuff" but rather "I love you guys!" The researchers used the familiar model of hazing to test the bounds of cognitive dissonance.

In one study, male undergrads were recruited for a straightforward task—a terribly, mind-numbingly straightforward task. They had twelve wooden spools and a tray with twelve slots. Using only one hand, they were asked to fill the tray with the spools, then remove them, then start again, for thirty minutes. But that was not all! When the half hour was up, the experimenter swapped trays, replacing the spools with a forty-eight-hole pegboard. As Festinger dryly described the subject's protocol: "His task was to turn each peg a quarter turn clockwise, then another quarter turn, and so on. He was told again to use one hand." In other words, subjects spent a miserable and pointless hour, and then—the icing on the cake— they were also asked to report to the next subjects: "It was very enjoyable, I had a lot of fun, I enjoyed myself, it was very interesting, it was intriguing, it was exciting." Whatever remote possibility there was of enjoying this repetitive nonsense task, there was truly no chance that anyone found it intriguing or exciting.

Now, the subjects were in fact offered compensation for their testimonials; they were told it was normally the job of a research assistant who was out that day. And here is the actual experiment: the subjects were all successfully cajoled into performing as spool-and-peg hype men—but half of them were promised $1 for that performance and half were promised $20. (This was 1959, so the $20 payment would have been shockingly high for a psychology experiment, about $180 in 2022 dollars.)

After they lauded the "fun" and "intriguing" task to the next group, the subjects were interviewed to give the department feedback on the "student experience" as psychology subjects. The main question of interest was "How enjoyable was this study?" The experimenters compared the $20 subjects with the $1 subjects.

Normally, I would think that people who get paid for something would like it more. It's fun making a lot of extra money! But that is not how cognitive dissonance works. Those in the $20 condition rated the task as less pleasant than the subjects in the $1 condition! The high payment let them admit that the task had been terrible, because they could still justify their behavior. Overall, the subjects had two dissonant thoughts: (1) *I just did a very dull task*; and (2) *I just told a bunch of people I loved it*. In the $20 condition, they could solve the equation by telling themselves, (3) *Well, sure, for $20 I'd tell a lot of white lies, including that one*. But those in the $1 condition had just wasted an hour and told a lie for not much money. How to resolve that tension? Change the belief: actually, spools are fun. Spools, pegs—what's not to like?

Overall, the studies on cognitive dissonance, from Festinger and from the research group he led, had a fiendish glee to them. Reading them today very much gives the sense that the Stanford psychology department in the 1950s and '60s was a freewheeling, bad-boys-of-academia sort of place, notably short of a strong institutional ethics review for human subjects research. (This is the same department that produced the Stanford Prison Study.) Elliot Aronson and another researcher, Judson Mills, were both graduate students in Festinger's lab when they wrote "The Effect of Severity of Initiation on Liking for a Group," a classic in the research on hazing and an artifact of its era. By way of framing, I want to note that this study is famous, relevant, and very, very sexist: it treats its female subjects like objects of prurient curiosity.

Female college students were invited to sign up for a discussion group on the psychology of sex. They were told that some people are too nervous to speak freely about sex, and that can ruin the group vibe, and so they were asked to take an "embarrassment test" to screen for participation in the sex discussion group. They were randomly assigned to two test conditions. In "severe embarrassment," they were asked to read aloud words that could be thought of as obscene (e.g., "cock," "screw"). In the "mild embarrassment" condition they read less vivid but still sex-related words (e.g., "virgin," "petting"*). After they read the words aloud, they were told that they had passed the test and could sit in on a session of the discussion group, although in fact they were listening to a pre-recorded conversation. They were asked to rate the quality of the discussion that they heard, and the subjects who had read the more embarrassing passages aloud were more enthusiastic about the discussion group. You can feel the internal pressure: if this was terrible, it had to be worth it. People can and will reorganize their thoughts to relieve the weight of feeling like a sucker.

Being a sucker is almost definitionally a state of cognitive dissonance: (1) This transaction is exploitative of me; and (2) I agreed to it. If those two statements feel bad, and the facts are stuck, it's the judgment that has to give way. This transaction is fine, not exploitative. This looks like a bad deal, but that's just how contracts are made these days. This fraternity pledge gauntlet was great bonding and all worth it.

Sometimes people will even report that the reason they liked

---

* Somehow worse than all the others, actually.

it was because they like being moral. In 2008, two psychologists (also at Stanford) wrote a paper they called "From Sucker to Saint." The researchers invited participants into a lab for a "study on communication" and told them that they would start the main study once the other participant, actually a confederate, arrived.

Subjects were assigned to one of three conditions. In the sucker condition, they were asked: Could you do a pretest for a different study on handwriting speed? The participants universally complied and wrote out numbers in longhand until they were told to stop. When the confederate arrived, he was also asked to do the handwriting task, but, within earshot of the subject, refused, saying he was pressed for time and it would be more expedient to get on with the main study. Subjects who had already been suckered into the handwriting task by the time they saw the rebellion rated themselves as significantly more moral than their partner. The authors concluded: "Just as people often rationalize their immorality, so they may sometimes moralize their violations of self-interested rationality—and thereby ennoble arduous endeavors that they might otherwise be inclined to avoid." Better to be morally superior than a dupe.

In this study, subjects were shown evidence that they had cooperated, for no reward, when others refused, and had to decide what to do with that information. What they did was to justify their choice—implicitly, at least—through self-righteousness: *I'm not a sucker, I'm a saint!* In some ways the switch reflects the mindset of QAnon supporters who defend their choices as an anti-child-sex-trafficking crusade, or the self-serving Trumpian rhetoric that claimed his political ambitions were protective of the American working class. Suckers are always in the eye of the beholder.

Cover

The essence of the peer-pressure cool-out is that other people exert pressure on your behavior. Sometimes the mark gets cooled out because they do the social math and realize it's easier to lick their wounds elsewhere. Those are situations where the sucker feels terrible but gets convinced to go home anyway. In some cases, though, the mark can be given a story about not being a sucker at all. It's not a firing; it's early retirement. It's not a demotion; it's just a different title.

Think for a moment of what it takes to perceive an outcome as a sucker's payoff: I have to feel like I cooperated, and I have to feel like my result is worse compared to someone else's, and that disparity has to feel unfair. But all of those constitutive judgments are themselves contingent and open to interpretation. Is it cooperation when I pay my taxes, or am I just doing what the state forces me to do? Is it unfair when I get assigned a lot of administrative committee work, or is it a compliment to my interpersonal skills? When the evidence is ambiguous, we are often compelled by the interpretation in line with our preferred outcome. This is the essence of motivated reasoning, that even when we are trying to work a problem through objectively, it is hard to ignore what we *want* the answer to be. The mind wants to settle on a story that gets the outcome we want without the feeling of being suckered.

To be more concrete: Imagine that I go to a new restaurant in Philadelphia. I am going to meet friends who are already there and whom I really want to see. The food also looks good, and I'm hungry. When I get to the restaurant, the host tells me that tonight the only option is a fixed-price dinner where you have to pay to get four courses. This is not the menu they described when I called in advance, and the total price is higher than I remember. However . . . I'm really hungry, and my friends are right inside looking very jovial and welcoming, and all in all it's not *that* expensive. Maybe I would feel like a sucker and still go in, or maybe I would feel like a sucker and cancel my plans on principle, but I think the most likely scenario is neither. I would probably shrug and think, *Eh, new management is just getting its sea legs,* or *I guess they changed their menu; that happens.* And voilà, now I get to eat my food and feel just fine about it.

Think back to the Ultimatum Game for a minute. In the Ultimatum Game, we have a pair of players attempting to divide $10 by, well, ultimatum. One player, the Proposer, makes a take-it-or-leave-it offer, and if the other player, the Receiver, leaves it, they both get nothing. What we know from years of research is that when the Proposer offers less than around 30 to 40 percent—$3 or $4 dollars out of $10—the Receiver will turn down the cash. It is a demonstration of what happens when the mark is *not* cooled out.

In a perfect world, what you'd like to be able to do as the Receiver is to take the money *and* feel good about it. But most people feel unable to take the $2; it feels too unfair and insulting.

What some researchers have leveraged, though, is that being offered $2 is not intrinsically unfair. We share our stuff at differential rates all the time; most of the time the people I meet share nothing with me and that's very normal. (I would find it odd if I met someone, even a friend, who announced, "I have ten dollars in my wallet, so I think the right thing to do is give you half.") The reason an 8-2 split feels bad in an Ultimatum Game is because there is no other context to explain the norms. With no other rules to guide behavior, most people fall back on a default rule of "fair" meaning a 50-50 split.

What if we start from the premise that the Receivers actually want to take the money—they want to take it and they are looking for a way to justify their choice? Some researchers have tested what happens when they could create a story for the players to accept the low offer and still save face. The descriptions can feel a little technical, but bear with me if you can stand it: these subtle nudges end up having this huge psychological and behavioral effect. Each study offers a little structural tweak that totally changes the sucker narrative.

The Uneven Split Study

Think for a minute of a regular version of the game where the players are dividing $10, but imagine that the Proposer is more limited. In the Uneven Split study, experimenters divide the pairs randomly into two groups. One group is told that, in this game,

Proposers can offer Receivers $8, $5, or $2. So those Proposers can basically offer a super-generous split, an even split, or a greedy split. The other group is instructed that the Proposer has only two choices: they can offer the Receiver $8, or $2. The question is: What happens when a Proposer offers $2?

Imagine that you are the Receiver in the first group. You know the Proposer could have offered you an even split, and instead you are getting offered just $2. As you might expect, those Receivers reject a lot of the time, more or less like what you see in a standard free-rein game. But now imagine you're the Receiver in the second group. You have been offered $2. You know your partner could only offer you an uneven split: it was either $2 for you or $2 for him. In that case, many more Receivers accepted the $2. Maybe they didn't prefer the smaller piece of pie—does anyone?—but the offer wasn't obviously exploitative or rude, because the Proposers didn't have the option of choosing an even split. For most people, the Proposer who has to choose between "unequal in my favor" and "unequal in your favor" is in a tough position, and it's not reasonable to require him to be *that* generous. It's not that unfair to get $2 in that case—and if the Receiver doesn't perceive the payoff as unfair, he can take the money and feel good about it. It's $2 either way, but in that second group, researchers gave the players a narrative for talking themselves into the better-than-nothing payout.

The Ambiguous Endowment Study

In the Uneven Split, the Receivers who got $2 knew how to calculate their share—$2 was 20 percent whether it felt reasonable or not. But another study produced a similar effect just by offering a plausible but uncertain story that said, *Maybe I'm not the sucker.*

The way the Ambiguous Endowment researchers did it was to run the game with nonstandard units of money, so players all got tokens instead of dollars. (The tokens had a fixed value—like, each one was worth a quarter—and they could be turned in at the end for cash.) Everyone was told that once the roles were assigned, the Proposers were going to be randomly allocated either thirty-eight or sixteen units. The Proposer would know how much he had, but, crucially, the Receiver would not.

The question for the research was clever: What did Receivers do with offers of eight or nine tokens? Think about the Receiver who gets an offer of eight tokens. That Receiver thinks, *Okay! Eight is half of sixteen; that's an even split, which would be a normal offer. Maybe the Proposer actually had thirty-eight to split, but I have no way of knowing. Chances are reasonable this is fair. I might not be a sucker. I'll take it.*

Now consider the Receiver who gets an offer of nine tokens. That Receiver thinks, *Huh, nine. What are the odds that my partner has offered me more than half of the tokens? Seems really unlikely. Which means that they must have been endowed with thirty-eight to share, which means they are offering me less than a quarter of the total.*

When they are offered nine tokens, it's a lot harder to tell a nice story. The end result was that the lower payout was in some cases more popular than the higher one. The story players told themselves when they got eight (*Maybe I got half!*) decreased the emotional cost of the low offer—no squawk or beef.

The Auction Study

The last cool-out version of the Ultimatum Game is the simplest, and maybe the most troubling. Before they started playing, participants learned that the game roles, Proposer and Receiver, would be allocated based on the results of an earlier auction. The Proposers were those who had won the earlier auction.

Receivers in the Auction game were more likely to accept low offers. If you think about it, you can see how they might rationalize their position. They have an offer of $2, say, and it would be nice to go home with a little extra little money—and they can tell themselves that the Proposer had earned his money, or won it, or deserved it.

It doesn't seem true in any deep normative way that the Proposer, by participation in some sort of nonsense auction task, had a true moral entitlement to the extra few dollars, but the auction let the Receivers save face, even just to themselves. They had an excuse, or at least a pretense, for allowing the transaction to go forward without disrupting their sense of self. At the extreme, accepting a low offer as fair means telling yourself you actually deserve the short end of the stick. That can be pernicious or it can be commonsense. Whether I'm offering myself good reasons or just easy rationalizations depends on context.

A Just World

One day early in my career I was listening to a talk by Omri Ben-Shahar, a contracts professor from the University of Chicago, who was describing the outrageous evolution of form contracts. He spun a fable of one hapless person, Chris Consumer, who would try to read all of his terms and conditions in a day and literally had no time to do anything else. To paraphrase: he couldn't make his toast because he was too busy reading the warning on the toaster plug; he couldn't enjoy his movie because he had to go search for the link that explained the limits of his rental license; he couldn't listen to music because he had to read the terms and conditions of Apple Music, and on and on. Like everyone in the audience, I was laughing—this is not that common in a law school faculty workshop so it is in itself memorable—but it was a turning point in my research.

For the decade since then, my research has been trying to answer the question: Why do people put up with this state of affairs? No human could possibly read all of their contract terms; it is literally not compatible with a normal economic and social life. I am a full-time professor of contract law, and I tell students every year that almost none of their contracts are worth reading. Yet courts and individual citizens alike are quite scoldy when it comes to complaints that someone has been duped into a deal whose devil was in the unread details. Popular accounts routinely describe even egregious forms of exploitation as the product of feckless consumers. (Unexpected banking fees? Illegal non-compete clauses in your employment contract? Your moving company *auctioned off your possessions?* Should have read the fine print!)

In my own research, I give people examples of harsh consumer deals and ask them to consider the implications. In one, for example, I described a car rental company that charged its renters triple if they received any parking tickets, even if the renters paid the tickets promptly. The scenario specified that the warning about the triple-fee penalty was very hard to find, written on an extra term sheet in the rental company's welcome folder.

I asked: Is it moral to require the customer to pay the fee? Is it a legitimate business practice? Did the customer really consent? I wrote the scenarios to be intentionally egregious, on shaky ground legally—but nonetheless most subjects thought it was all right. They seemed to think that consumers like themselves sometimes consent to bad deals, but the bad deals are still fair play. No suckers here, just rational risk-takers winning some and losing some.

In the Ultimatum Game, players are motivated to take the money that they are offered; more money is usually preferable to less. My research kept leading me to the sense that people feel *motivated* to see consent in places where it is clearly compromised, but this did seem odd. Why would research subjects, none of whom are titans of business themselves—if they were, they would not be taking online surveys—be rooting for the companies rather than the little guy?

There are enormous psychological consequences to believing that the world is unfair—it feels really destabilizing and depressing—and people will adapt their other beliefs to fit a story about a just universe. The feeling of living in a rigged or random system can feel terrifying and unsettling in a way that is worse than losing and feeling like you deserved to lose. People accept oppressive contracts from their corporate counterparties for the same reason they accept other kinds of banal exploitation from institutions: it just feels better to justify an intractable situation than to grind against it.

The motivation to justify the status quo is strong enough to excuse a lot of bad behavior by the powerful, even in the universally maligned world of fine-print contracts. Uriel Haran, a professor of management at Ben-Gurion University in Israel, wrote a remarkable paper about what happens when companies break their deals. He had asked subjects whether a breach of contract is just as wrong for corporations as it is for individuals. They reported that it is wrong for individuals to breach their deals—but not that bad for companies.

The picture that these studies paint is one that gives corporations, but not individuals, the OK to hide their bad terms, or to draft reasonable terms and then violate them. It seems perverse; surely big companies should at the very least be held to the same ethical standards as regular people. (I'd argue we should cut them less slack, not more, given their resources and power.) When parties in power take advantage, it is not coded as a sucker's game. It is not coded that way because then we would have to live with the feeling of being a sucker all the time, and that would feel terrible.

In the 1950s, a heyday for these grand psychological theories of social life, the psychologist Melvin Lerner was grappling with this same question of motivation: What would make people so *up* for being cooled out? Lerner's theory was called the just-world hypothesis. He argued that people are strongly motivated to feel like they are living in a just and orderly universe. For Lerner, this bias explained something deeply troubling that he observed in his life and in his laboratory: a refusal to reckon with the complexity and unfairness of the real world. Lerner developed a theory that a "just-world bias" affected people's judgments, skewing their beliefs to favor evidence that the world is structurally fair.

In the original experiments, Lerner and his coauthors would invite groups of college students to participate in a task that claimed to measure individual stress responses. One of the "subjects"—someone actually working with the experimenters as a confederate—would be chosen to receive shocks while completing a memory task. (The shocks were fake.) This person, supposedly trying to learn in stressful conditions, was shown on CCTV to the rest of the subjects, who then answered a battery of questions about the learner's stress adaptation.

What Lerner wanted to know was how the subjects would justify the fact that one of their own group members had been randomly selected to get bad shocks. He assigned the subjects to one of two conditions. For some, they had the power to stop their classmate's shocks halfway through. Others just had to watch powerlessly. Lerner's prediction was that helpless observers with a "just-world bias" would blame the victim; when you see something bad happening and can't stop it, you can tell yourself a story about how it is actually *good* or *right*. Indeed, the subjects who could only observe, and could not help, rated the victim as being a worse person—less attractive, of poorer character. Stuck watching what appeared to be a difficult and random victimization of a peer, subjects implicitly decided that she just deserved it. Subjects who could intervene were less committed to a narrative of shock desert.

Lerner's book, *The Belief in a Just World*, was subtitled *A Fundamental Delusion*: he thought people's need to find moral order in their world was so strong and resistant to evidence that it was actually delusional. We judge others as deserving of their bad luck, and we judge ourselves this way, too. Victim blaming extends to the self, even when it seems odd that anyone would be psychologically motivated to accept blame for their own victimization.

The psychologists John Jost and Mahzarin Banaji took just-world theory into the twenty-first century with a set of experiments and hypotheses that looked more broadly at the motivated reasoning behind victim blaming. They argued that the motivation isn't just to believe good things happen to good people, but about something one level more sophisticated: the drive to justify the system. Some justifications are made at the expense of oneself and one's community in order to preserve the "existing social arrangements"—otherwise known as the "system." Writing separately, Jost described his research agenda as asking "Why Men (and Women) Do and Do Not Rebel." Whether the system is the patriarchy or the meritocracy or even the law, these are features of society, he argued, that are so fixed that rebellion feels futile and demoralizing. This is what Goffman was talking about when he wrote about social sanitation, the social norm of "persuading persons to keep their chins up and make the best of it . . . enjoining torn and tattered persons to keep themselves packaged up."

In the final reckoning, the cool-out itself is sometimes just another sucker's game. Large-scale scams have been successfully perpetuated by accusing doubters that the very suspicion that they have voiced is itself evidence of being duped by the Man. Donald Trump, a scammer par excellence, made his political name with relentless exhortations that his followers were being tricked by everyone from the media to the Clintons to asylum seekers at the border, even as his hapless, vicious presidency made him richer every day. Fox News makes its money by peddling fear, with paranoid claims that liberals are sneaking critical race theory into schools or lying about COVID-19 vaccines. In the meantime, Fox News is itself a multibillion-dollar grift—or that is certainly the argument of reporter Brian Stelter, who quotes a producer as saying, "We don't really believe all this stuff. We just tell other people to believe it."

And, of course, QAnon is an elaborate hoax. An anonymous poster in internet chat rooms claims he is a government insider with special knowledge of a vast conspiracy of child sexual exploitation involving liberal elites. All evidence suggests that Q has neither the knowledge nor the access that he claims; that he is making up his evidence, clues, and warnings; and that there is no abusive cabal. Many, many people see QAnon for what it is: a malicious scam. But for those who have gotten into it, it is hard to get out. Evidence stops mattering. You don't think Joe Biden is an agent of the Deep State? Don't believe what they tell you. Do your own research. That's how they get you.

CHAPTER 8

Mothersucker

I was a twenty-eight-year-old graduate student when I had my first child. My pregnancy was public at an early stage; my unborn child grew perpendicular to my body. By the halfway mark, strangers would chat to me about my distended abdomen and inquire about twins or imminent labor. People sometimes offered me their seat on the bus or a bottle of water. It was overall both convivial and intrusive in ways that felt familiar to me as a person who had been living in public with a woman's body since middle school.

I was surprised and humbled to then have the kind of birth—surgical, emergent, laced with terror—that you try not to think about once it's over. After some weeks of hazy, tense recovery, I put my oversized newborn into a BabyBjörn and went out into the world. I took him with me when I was getting coffee or groceries or when I picked up mail or papers from my shared office at the university.

Exhausted, baby in tow, I discovered that things had changed during my confinement, or maybe I had changed. In little ways (bus seats, eye contact) and bigger ones (unpaid leave, day care waitlists), it was clear: as a mother, my stock had plummeted.

The writer and political commentator Ann Crittenden recalls a dawning epiphany in the early 1980s when her child was an infant that mothers have "literally, the most important job in the world." A quick google will give you a sense of the Hallmark and Etsy agreement around this sentiment. Then she had a very different epiphany.

"I'll never forget the moment I realized that almost no one else agreed," she wrote. "It was at a Washington, D.C., cocktail party, when someone asked, 'What do you do?' I replied that I was a new mother, and they promptly vanished. I was the same person this stranger might have found worthwhile had I said I was a foreign correspondent for *Newsweek*, a financial reporter for the *New York Times*, or a Pulitzer Prize nominee, all of which had been true. But as a mother," she concluded, "I had shed status like the skin off a snake."

The theme of this chapter may feel unusually specific, and its tone and content are certainly more personal. But the argument here is a natural culmination of the broader case I have been making in this book. The false promises about motherhood, and the myriad of sucker stories about women bearing children, are deployed to reinforce traditional social hierarchies that depend on the subordination of women and people of color. In the first seven chapters, I covered a lot of ground: the ways that people feel duped, or feel cooled out; the ways that people avoid or react to the prospect of being suckered; and the ways that sucker themes perpetuate and define racism and sexism. In surprising places, motherhood is a fool's exemplar, reflecting and even magnifying themes of power, status, duplicity, and morality.

The social construction of motherhood is a deep bait and switch. Motherhood gets a lot of good press, but women caring for children are perceived, across the board, as being of lower status than other women, and lower status than men of any caregiving situation. On its face, becoming a mother is a cultural achievement, a self-actualization that secures one's place in the community. In fact, motherhood comes with heavy responsibilities but not a lot of power. Supermoms get Mother's Day cards but little material support.

The promise of motherhood in the abstract is a promise of love, yes, but also of status. Moms are American cultural icons; they rock the cradle and rule the world; they know best; they are heroes. These are the aphorisms announcing high status—moms associated with patriotism, power, competence, and valor. Mothers get a lot of reverential lip service (literally, even: "You talk to your mother with that mouth?"). The cultural narrative of mothers suggests, if superficially, that achieving motherhood has recognized and rewarded social value—at least for white women married to men. The depth and richness of the intrinsic rewards of motherhood are real, but the social and political rewards are much less reliable. This chapter has a perspective, of course, and it is very personal. But the only personal experience I have of motherhood is my own and, in terms of structural support, my demographic is sort of a best-case scenario. White married women like me get the *most* credit, and the most backup, for our maternal roles.

Women's value has always been measured in childbearing. "The Clock Is Ticking for the Career Woman," wrote Richard Cohen when he coined the term "biological clock," an expression that conjures nothing so much as a woman's body counting down to its own expiration date. If social rewards are the carrot, they are tied up with the stick. Motherhood is more quietly enforced with a threat: no social promotion without reproduction. In the words of my colleague, the sociologist and law professor Dorothy Roberts: "Society, at one level or another, exerts structural and ideological pressures upon women to become mothers."

Motherhood implicates a web of implicit bargains. We have transactional relationships with the state, with co-parents or romantic partners, with our children, and with our communities. In each of these relationships, women bearing and raising children create public goods. I have two children, and they are, indeed, the organizing principle and the core fact of my adult life. But it is an area of my life where I understand a distinct social norm, sucker-wise. In a culture of otherwise vigilant sugrophobia, moms are invited, indeed encouraged, to look the other way, to overlook the possibility that they are getting the short end of the stick.

On reconsideration, my first clue about motherhood should have been the birth itself, or if not the birth, the fact that I wasn't supposed to complain about it. I don't know how bad my son's arrival was on a normal distribution of births; I know for sure it is the most physical terror and the most medical danger I personally have ever been in. Labor ground on fruitlessly for a night, a day, another night; I spiked a high fever; they called for an emergency C-section; I lost pints of blood while my husband turned mute and gray by the operating table. They whisked the baby away.

After a deep, black morphine sleep, I awoke, and the world had moved on. Doctors and nurses adopted the same no-eye-contact good cheer when they came into the room, calling me "Mom," reminding me to "sleep when the baby sleeps." Reeling, anemic, I took my son home as soon as they would let me out. (In my follow-up appointment, the ob-gyn from my practice asked about planning for the next child. I gently pointed out that the birth they had attended was nightmarish, was still giving me actual nightmares; it would not be sensible for us to move forward together. She was stunned and defensive, like what did I *expect*? Birth is *hard*. I was surprised back. Birth is hard, yes, but I'm still a rational consumer.)

I had gotten the memo from an early age: being a mother is what society respects. If I had been asked directly, I think I would have said that I wanted to have a child (very true) and also that I understood my choice to be the fulfillment of an expectation. It is not that I chose parenthood in order to procure social rewards—or at least I don't think so—but I did think I was complying with something basically socially normative. From the outside, motherhood looked to be a prerequisite for social status for adult women. Certainly, women without children get the message that their choice is suspicious, if not pathetic.

In any case, my physical transition to parenthood—i.e., childbirth—left me feeling foolish and naïve. People with high status get their painful experiences taken very, very seriously. Their suffering is remediated; their foibles are tolerated; their problems are solved. As the mother of a real, live baby of my own, I could see that I had deeply misunderstood the deal.

Devoted

At the basic biological-anthropological level, mothers are producing the next generation of humans. The cost of bringing children into the world is high. Pregnancy is an enormous investment and physical commitment; childbirth is painful and risky; women are often feeding children with their own bodies. The ideal mother does the work of reproduction and care but doesn't call it work, or in any case doesn't complain about it. "Patriarchy would seem to require, not only that women shall assume the major burden of pain and self-denial for the furtherance of the species, but that a majority of that species—women—shall remain essentially unquestioning and unenlightened," wrote the feminist poet and essayist Adrienne Rich. We rely on and name a lot of selfless mothers—the nursing mother, the stay-at-home mom, the Supermom, the class mom, the soccer mom, the snack mom, even the mom-to-be. Traveling in France with her family, Rich, mother of three sons, wryly took stock of one affectionate jab. *"Vous travaillez pour l'armée, madame?"*—"Do you work for the army, ma'am?"

The attributions of selfless devotion themselves contain a little hustle within. They seem so good! Superman, the military—those guys get lots of respect! Breast is best! And everyone surely loves Snack Mom and appreciates the class mom. And yet . . . you can feel the undertone, the sense that you might be thanked but not respected. When he was writing about the cool-out, Erving Goffman observed, "Sometimes the mark is 'kicked upstairs' and given a courtesy status such as 'Vice President.' In the game of social roles, transfer up, down, or away may all be consolation prizes." He helps put language to just what is so discomfiting about Supermom. It is an accolade of demotion, a prize for the mark, not the operator.

It reminded me of an episode of *The Wire* from 2008. An editor and a reporter are smoking outside the newsroom, shaking their heads over reports that a mother of four has died from an allergic reaction to blue crab. The editor, a laconic veteran of the headlines, says, "You ever notice how mother of four is always catching hell? Murder, hit-and-run, burned up in row-house fire, swindled by bigamists. Tough gig, mother of four."

Cynthia Lee Starnes, a law professor at Michigan State, was writing about divorce settlements when she apprehended the mischief that notions of selflessness were wreaking in the courts:

> Many mothers have been stunned to learn that after years of viewing themselves as proud and valuable contributors to marriage, to family, to a new generation, the law of divorce views them as suckers. Surely this is a mistake, a mother might insist, a confusion of identities, a dialectical lapse that will be corrected as soon as it is discovered. Sadly, there is no mistake. The dispiriting message is that primary caretakers, the vast majority of whom are mothers, have been duped into providing free family caretaking at great personal economic cost; a price they must pay for their imprudent ways.

In various and unexpected moments, even divorce court, mothers get these little hits of disrespect, when a tableau sharpens briefly. A person blithely assuming her work is respected, her role valued, has a gradual realization: "Ohhh, wait . . . you all think *I'm* the fool. . . ."

It may be counterintuitive to think about mothers as low-status, but the evidence bears this out. In fact, attributions of foolishness—incompetence—are endemic to motherhood, anecdotally and empirically. In the stereotype content grid, "homemakers" are categorized as high-warmth but low-competence.

In fact, in a separate study, that same research team decided to follow up. Is it motherhood itself that causes attributions of incompetence? They knew from their earlier surveys that homemakers were viewed as low-status allies, "liked but disrespected." But it was hard to know what facts caused this attitude. Maybe lack of respect for homemakers has to do with beliefs about who opts out of the labor force: they couldn't cut it, they are not ambitious, and so on. Or, instead, maybe the stereotypes are triggered by the mere fact of motherhood. This is what they set out to test.

The research team asked 122 Princeton undergraduates to do a short person-perception task with the prompt: "We'd like you to read the profiles of three consultants at McKinsey & Company's Manhattan office and give us your first impressions of them. Imagine you're a client, trying to choose a consultant from very little information." Two of the profiles were just filler material, to help obscure the researchers' real goal. The target profile described either Kate or Dan, a telecommuting worker from New Jersey. In the version with Kate's name, it read:

> Kate is a 32-year-old associate consultant who graduated with an MBA. She's been working in her current field for six years. When working with a client, her duties include identifying issues, planning and conducting interviews and analyses, synthesizing conclusions into recommendations, and helping to implement change in her client's organizations. Her hobbies include swimming and tennis. Kate and her husband recently had their first baby. She lives in central New Jersey, commuting to work two days a week and telecommuting three days a week.

This same vignette was distributed in four versions: Female/Mother (above); Female/No Kids; Male/Father; Male/No Kids. If the vignette was about a male, the pronouns were masculine and the target's name was Dan. The No Kids versions simply omitted the penultimate sentence.

Subjects were asked to rate each profile on a series of traits related to competence (capable, efficient, skillful) and warmth (good-natured, sincere, trustworthy) and also answered questions about whether they would want to hire and promote the consultant.

Men and women without kids were both rated somewhat higher on the competence measures than on the warmth measures. This makes sense, given what we know about the broad category of "professionals" in American life. Whether male or female, parents or not, members of the professional class tend to elicit some envy and distrust. This bears out in the stereotyping studies, but it is also in line with what we would informally think of as class resentment. Professionals at a baseline are viewed as competent but not that warm. But when male professionals were described as new fathers, their competence measure remained steady, while their warmth measure increased. Okay, that's reasonable. The job is evidence of professional competence, and the addition of the baby is a little humanizing or softening.

For women, though, that softening came at a price. The new mother was rated as significantly warmer than her childless self—but she was correspondingly downgraded in competence. The fact of motherhood made her less capable and efficient.

We might be tempted to think that this perception could still cash out to something all right for professional women—that maybe for women to have status, what is professionally important is that they are perceived as warm. (I guess I am not personally tempted to be okay with this, but I can see why we might ask.) However, Cuddy, Fiske, and Glick also asked subjects directly about professional success. For men and for women, it was competence, not warmth, that was associated with "positive behavioral intentions": subjects were more interested in hiring, promoting, and educating the childless woman, or the childless man, or the working father, than the working mother. That downgrade in perceived competence was costly.

It's costly in smaller, funnier ways, too. I recently scanned a set of parenting headlines roughly marketed at my demographic. There was an article that might help me capture my "Mom Boss Energy"; a cover story titled "The Juggle Is Real but You've Got This"; and a feature called "Mom Inspo for the Win," among others. Periodically I am interacting with the world in my capacity as someone's mother and get a line of feedback or direction that suggests that I am a child myself rather than the parent of one. I once took my daughter to the pediatrician with a purplish rash on her hand that had persisted for about six weeks. The doctor looked at it for a few seconds and then looked up, almost concerned for *me*, and said it was probably Magic Marker or maybe a Sharpie. I imagine that this kind of mistake happens to parents; I have seen enough episodes of *House* to know that doctors are supposed to look for horses and not zebras, etc. Nonetheless, I looked back at her, temporarily stupefied. I am not a rocket scientist, but in most settings I am taken for the kind of person who can diagnose Toddler Drew on Herself without medical advice.

One of the tropes that fascinated me when my kids were little, and then when they were older, but with darker undertones, was an odd through line of suggestions that my children themselves were tricking me. Aside from our truly harrowing encounters with pediatric skepticism during more serious illness, the narrative seemed to be built into everyday parenting advice. Some of the popular sleep training manuals suggested parents should watch out for babies who cry to manipulate the sleep schedule, or even teach themselves to vomit to get attention at bedtime. It riffs, of course, on a kernel of something all parents experience: the intermittent sense that your children might be master puppeteers, somehow able to time a wail to coincide with the first moment of REM sleep you've had in days. At the same time, it suggests a role reversal. Am I really afraid of being tricked by a child? What kind of a fool do you think I am?

Demoted

In my personal day-to-day, I do standard mom stuff on my kids' agendas. I'm all in. I will tape lollipops to sixty little red cards on February 13 and find an outfit for each day of Spirit Week. (Dress like your teacher! Sixers Day!) I will drive to South Jersey every sweltering weekend in July so my kid can play under-eight travel baseball. (For the record, there is so, so much crying in 8U baseball.)

As a commentator on a broader set of cultural norms and standards, I also observe that some facets of the maternal experience can seem closer to hazing, or false consciousness. I remember a mild bemusement when I was first pregnant and interested in learning more about evidence-based obstetric care. When the doctor gave me a list of things not to ingest, I was curious. What are the risks of sushi? What harms does NyQuil cause a fetus? Are there statistics about unpasteurized cheese I could have a look at? Hahaha, no. These things could have pathogens or chemicals that could, in theory, be harmful in high doses. There are some reports of bad outcomes. There are no randomized controlled trials to estimate the level of harm. (I didn't consume them! I'm basically a rule follower!) My idea that doctors would need some sort of cost-benefit analysis to justify their recommendations—what level of risk justifies reducing my dinner choices?—was not the methodological norm. Any probabilistic harm to the baby counted as a real cost, whereas, it became clear, my experience of pleasure or relief was my own business.

There is a *Calvin and Hobbes* cartoon in which Calvin is negotiating with his neighbor, Susie, whom he has asked to dare him to eat a worm. She has provisionally agreed, and the question is whether he is going to be paid up front or only on completion. She tells him it's on completion or no deal, and he grumbles, "Man, you'd think the guy eating the worms would be calling the shots."

"Usually, if you're calling any shots at all, you're not eating worms," she replies.

Which brings me to the breast pump. I didn't have to consider the breast pump in earnest until my daughter was born. I had had my first kid when I was a graduate student, and I made my schedule around his schedule, because I could and also because I didn't have a very good plan going into the whole thing. We kept that up for his first year while he cheerfully yet resolutely declined every bottle offered to him. I had a different set of constraints with the second baby. The timing of her arrival was academic perfection, born at the end of April right as students were finishing their final exams. I returned to teach the following September with a robust four-month-old who was accustomed to eating every few hours. When I took a borrowed breast pump into my office—my private office! With a lock on the door! It was literally the best-case pumping scenario!—to see how this whole system might work for us, I caught a glimpse of what I can only describe as deep existential disrespect.

It's not that I didn't know many other women—most of my friends, for sure—who pumped for months from their offices or cars or even office bathrooms, who found it reasonable enough: just another piece of household business that they folded into their schedules. I definitely get that breast milk is nutritious for babies, and I am otherwise someone who will be medium fussy about giving my kids Red Dye No. 40 or Cool Ranch Doritos. But I could not get past the sense that it was objectively outrageous, that no one would ever, ever ask my husband or my male colleagues to do anything like pumping. The prospect of taking my shirt off at work, hooking up a milking machine to my body, maybe opening up my email so I could use my time productively while the pump whooshed on, and finally sanitizing that machine part by part? But discreetly because people find it kind of gross? Ah, I see, I'm eating the worms. (Reader, I switched to formula.)

My point may feel frivolous, like a dated blog rant, but it is emblematic of something deeper. There is an abasement, an infantilization, to getting down to a child's level, about agreeing to the child's agenda, whether the agenda is rainy day crafts or breastfeeding. Nursing a baby, a substantial public and private good, nonetheless makes the mother, in public at least, the stooge.

Indeed, three researchers at Montana State found this to be the case. When men's attention was drawn to the fact that a woman was breastfeeding, even out of sight, they perceived her as less competent at work and, oddly, worse at math. It is not that there is a truth of the matter—I don't know the truth of the matter—but more that it is part of the bait and switch. Indeed, breastfeeding alone is the subject of a relentless campaign, an Unequivocal Good—but not a function that brings with it attributions of skill or status. Worse, with it comes assumptions about the mother's professional incompetence. You can be a good mother or a respected professional, but not both.

"A mother's work is never done," it's true, but there is a shared sense of a social pact, that the labor and commitment is part of a package deal; it is work in return for both the joys and love of raising children along with some acknowledged social respect. But motherhood can be like playing a relentless Public Goods Game in which your job is to contribute, but everyone else gets to play whatever strategy they want.

In a *Saturday Night Live* skit from the bleak, bleak winter of 2020, Kristen Wiig starred in a frenetic Christmas morning musical number as the weary but game mother of a small family. Her kids and husband are exclaiming over their gifts on beat ("I got a phone!" "I got a watch!" "We got the piano from *Big*!"), and as they go around and around with new gifts, stocking gifts, hidden gifts, the camera periodically returns to the mother, who says, "I . . . got . . . a robe." At one point, one of the children makes meaningful eye contact with her mother and says, "Looks like someone's got a big surprise . . . it's presents for THE DOG!" (Sister: "So many presents, but he deserves it!") At one point, as the children remove ever more elaborate tech toys, the mother says, "I'm gonna make us breakfast."

Doing It Wrong

Being a parent is extraordinarily demanding, to state the obvious. It's relentless in a way that most market labor is not, and you can't opt out. But it is also subject to a level of scrutiny, not just for making moms worse at working, or math, or being hot, but for doing the job of mothering itself wrong.

When my kids were little, I was put in mind of the claw game at an arcade: it looks like all you have to do is play the game and you win something, but of course it turns out the claw is purposely impossible to use and you'd have to be some sort of arcade savant to get a prize. Even if there is a model mother who gets universal accolades, the odds that you personally are going to achieve that status are very low. There are canonical mothers in popular culture, characters who are the moral center of their universe and receive respect from their children, spouses, and communities: Marmee March, June Cleaver, Clair Huxtable, Tami Taylor. But the dignified matriarchy is not that easy to come by.

Social approval is reserved for mothers who have successfully navigated a gauntlet of grueling choices. Did they "lose the baby weight," breastfeed "on demand," establish a bedtime routine, schedule regular date nights? The mismatch between the cultural conversation (stay hydrated!) and the deep meaning (the work is nonnegotiable and unpaid) is so absurd that its best expository medium is the satire. There is a Twitter feed called Man Who Has it All, a character who purports to offer advice to other men who want to Have It All, as in, "Working dad? Guilt is your worst enemy. Tackle guilt by staying up late to get all your housework done and getting up early to finish it off." The send-up hits all the points: the combination of structural and physical stressors with punishing "solutions" and inane pick-me-ups, the notion that the negative response itself is something to be managed on your own. "Is it REALLY possible for men to juggle housework, career, 2 or 3 almonds, a naturally radiant posterior, and 'me time'?"

In the real world, of course, it is women rather than men who see their parenting policed at the level of minutiae. The clichéd reports are also empirically true: mothers are held to a higher standard of parenting than fathers. Recently, psychologists wanted to study perceptions of "parenting effectiveness" in a world of shifting gender roles. A research team gave undergraduate subjects, both men and women, a scenario in which they described a parent, either Lisa or Gary, juggling work and home life. The version with Lisa read:

> Lisa is a 30 year old, married woman with a two year old child. Both Lisa and her husband are employed full time as newspaper reporters and work outside the home from 9:00 to 5:00. Lisa worked as a reporter before her child was born and then resumed working full time at the end of her six-week parental leave. Lisa's primary reason for working is that her career is financially beneficial to her family; that is, her career helps her family maintain an acceptable standard of living.

Subjects were randomly assigned to read the same facts about a mother in this role (Lisa) or a father (Gary). They then got a list of eleven parenting behaviors: bathing, preparing meals, comforting, reading together, playing, and so on. They were asked to estimate the number of times per week or hours per week Lisa or Gary would be doing this kind of parenting. At the end, the subjects also answered a more omnibus question: "Overall, what kind of a parent do you think Lisa/Gary is?" They responded on a scale from 1 (very bad) to 7 (very good).

The scenario varied the parent's gender and also varied the parent's working status: some versions of the vignette had Lisa or Gary as stay-at-home parents. The results were telling, and the story they told was that the mothers could not win. Whether Lisa was working or staying home, subjects thought she was doing more physical and emotional caregiving work. Okay! Seems like praise! It was, but only sort of. When stay-at-home Lisa was credited with more caregiving and stay-at-home Gary was credited with less, they were still rated the same. When working Lisa did more work than working Gary, she was judged a worse parent than him on the all-around score. All of that credit added up to very little purchasing power, respect-wise.

These may seem like marginal judgments, but maternal failure is a really easy, salient explanation for a range of social ills. Did you feed your baby formula? (Obesity.) Are you a helicopter mom? (No life skills.) Did you coddle them? (No work ethic.) Historically, a course in "abnormal psychology"—i.e., psychopathology—might take semi-seriously the possibility that it was the mothers who were harming their children. Autism was blamed on the "refrigerator mothers," who were not warm enough to encourage emotional development. Eating disorders were blamed on the "enmeshed" mothers, who were *too* warm—psychologically too close for developmental individuation. Mothers desperate for attention might cause their children to have a factitious medical disorder by Munchausen syndrome by proxy. There is a reason that the go-to psychoanalytic trope is "Tell me about your mother." Moms! They're always doing it wrong!

There have been times in our parenting lives when my husband and I have purposely swapped roles, giving him a recalcitrant toddler in an airport or a challenging parent-teacher conference not because it was convenient or preferable but because we knew the particular public would just cut him more slack. A dad at the playground who runs out of snacks, who asks for extra coverage to bring one kid to the bathroom while the other keeps playing—what a dad! So involved! A mom in that situation is strange, struggling, presumptuous.

Draconian judgments of imperfect mothers inhere even at the most horrifying margins. Two social psychologists wrote an article called "Mother Knows Best so Mother Fails Most" that focused on the "punishment of parenting mistakes." They asked subjects in their study to play the role of jurors in a criminal trial for a parent who made the fatal mistake of leaving a toddler in a hot car. (I can barely stand to even write these words.) The literal nightmare scenario was run with either a male or female parent as the defendant, and the findings were clear: all subjects rated a woman who made this mistake as overall worse at parenting than a man in the same situation, and male subjects in particular imposed longer prison sentences for female defendants than male defendants.

These cultural parenting questions—who gets policed, who gets the benefit of the doubt, who gets credit, who gets blame—are even starker at the intersection of gender, parenting status, and race. For Black and brown women, the tightrope associated with mothering is a thinner thread suspended over a deeper ravine. Black women especially are policed with a viciousness that leaves no room for error, or even judgment. Black mothers are monitored by child welfare agencies and separated from their children at dramatically disproportionate rates. Sometimes their parenting is literally policed, in the sense of criminal consequences. We have seen this already: the parents who allowed their child to relieve himself alongside the car on a trip and were arrested in Ferguson, Missouri, or the mothers jailed for changing addresses to get their children into better schools. An Ohio mother of three, living in a motel with her children, was arrested when she left her nine- and three-year-old daughters in the room while she worked at Little Caesars. In South Carolina, a Black mother was arrested for letting her nine-year-old play in a nearby park while she finished her shift at McDonald's.

The cultural recruiting materials promise that being a mother is the best and most important job in the world, which in some ways it is: it is a singular experience of love. Once you get the job, though, the *social* rewards remain just out of reach, unless you're married, white—and never, ever make a mistake.

Mom Con

In the United States, motherhood is a social good until women ask for state support, at which point motherhood is a personal choice.

When I was in law school, I took a very formal class on law and rationality. We read political and legal philosophy texts that I found quite abstruse, and I could barely keep up week to week. So it was a surprise to me to read a passage in one article, assigned for a unit on the theory of the social contract, that generated not just comprehension but outrage.

The philosopher David Gauthier had written about the mutual obligation of members of a society to try to be net contributors, and he needed a salient narrative to bring the theory to life. This is what Gauthier chose:

> Let me illustrate the application of this requirement with a deliberately provocative example. Consider women on welfare who choose to bear a child—or who, becoming pregnant, choose not to have an abortion. Such women have an unfavorable image these days and, on a contractarian view, this may seem well deserved. In choosing to bring children into the world despite their lack of both financial and, in many cases, emotional resources to care properly for those children, they are violating an appropriate extension of the proviso that forbids interactions that better one party by making the other party worse off than anyone need have been.

Looking to illustrate, in a familiar way, how a social exchange might be exploitative, he landed on mothers—not as the dupes but as the exploiters. (Many thanks to the guy in my seminar who, called on to respond to this passage, said, "What if they are, like, good at being mothers?")

Mothers come under scrutiny when they are suspected of putting one over on society. Specifically, they might become mothers for the "wrong" reason: to extract benefits for themselves rather than to love and nurture the next generation.

The skepticism is especially trenchant for mothers who are poor, born outside of the United States, Black, brown, indigenous, and marginalized. As Dorothy Roberts wrote, "Underlying the current campaign against poor single mothers is the image of the lazy welfare mother who breeds children at the expense of taxpayers in order to increase the amount of her welfare check. In society's mind, that mother is Black."

Much like the contrast between benevolent and hostile sexism, when it comes to mothers there is condescending reverence on the one hand and a concomitant fear, on the other, that women will use their reproduction to take advantage. On this view, having children gives women leverage; it is a tool they might use to extract everything from better shifts at work to American citizenship. They might use pregnancy to trap a man. They might use paternity claims to procure child support. They might have children just for the extra welfare benefits. In other words, they might cynically use the moral imperative of motherhood to sucker others—men, the state, their coworkers—into paying their way.

For many years, the rhetoric of welfare has been a set of warnings about women, especially Black women, scheming against the working righteous. In a sociology of welfare benefits, for example, one African American woman paying with food stamps recounted, "When you get ready to buy your groceries, people have made nasty little remarks about the groceries you're buying. They'll go, 'We're paying for that.'"

Charles Murray, author of *The Bell Curve*, made the case plainly in 1994. He wrote a paper entitled "Does Welfare Bring More Births?" His proposition was that welfare causes motherhood: women were having children as a means to obtain welfare benefits. He mused:

> I find it absolutely fascinating that illegitimacy is so intensely concentrated among very low-income women. I'll give you an example. White women—don't talk about blacks—among white women who are below the poverty line in the year prior to birth, about 44 percent of all their children are out of wedlock. White women anywhere above the poverty line in the year prior to birth, it's 6 percent of births. Now, come on. There's

got to be some bizarre reason why very poor women suddenly find it desirable to have babies out of wedlock, whereas not poor women don't.

In other words, women who are eligible for welfare benefits use their reproductive capacity to scam the system.

Coming from Murray, widely derided for relentlessly "peddling junk science about race and IQ," it is easy to dismiss as a fringe view, but in fact he was saying plainly what others implicitly endorsed. In 1996, President Clinton's welfare reform permitted states to impose family caps, and the rhetoric was not dissimilar. The government would not incentivize more children for poor mothers—suggesting a shared belief that poor mothers were having extra children to work the system.

Indeed, if we think that mothers are always navigating a fine line, we may think about Black and brown mothers as working against a presumption not only of incompetence but also of malice. As Dorothy Roberts wrote, "American culture reveres no Black madonna; it upholds no popular image of a Black mother nurturing her child." She argued, more starkly, that the mothers most likely to be accused of scams are the ones who are not producing white babies: "White childbearing is generally thought to be a beneficial activity: it brings personal joy and allows the nation to flourish. Black reproduction, on the other hand, is treated as a form of degeneracy. Black mothers are seen to corrupt the reproduction process at every stage."

Roberts points out a fundamental race divide in the maternal transaction. White mothers are patronized and supervised, and their access to the rewards of motherhood is contingent on their docile acceptance of a mediocre deal. But if white motherhood "allows the nation to flourish," then even if those mothers are using food stamps, or leaving work early, or lying about paternity, their con is relatively benign. The scam looks much more threatening from women whose contribution to the social good—raising Black children—is itself devalued or ignored.

In some ways the perceived threat of non-white motherhood has been laid out with unusual candor in the context of immigrant women, especially women from Latin America. For Latina mothers, reproduction and fertility are associated with the dark warning that their reproduction is a plot for undeserved citizenship and, ultimately, a massive demographic shift. The pejorative term "anchor baby" refers specifically to the child of an undocumented woman who has used her childbearing capacity to extract the benefit of citizenship for a whole family not otherwise eligible for legal immigration. In 1994, when California voted on a bill to deny social and health services to undocumented immigrants, one of its advocates made the plain case that if you permit the state to offer benefits to undocumented women, they will use their fertility to increase access to undeserved entitlements. "They come here, they have their babies, and after that they become citizens and all those children use social services," she explained.

Implicit in these claims is the threat of Latina motherhood as a plot for white replacement. Indeed, the right-wing "replacement theory" of demographic change accuses women of maternal malfeasance across the board. White women reproduce too little, failing to hold up their part of the racial bargain; brown women reproduce too much, threatening the status quo. Merely by the fact of their ethnic heritage, Latina mothers are subject to racist skepticism that they are not "real Americans," no matter how deep their family roots in this country. The anthropologist Leo Chavez has described the "singling out [of] Latinas and their 'fertility problem' as the cause of negative demographic changes (proportionally fewer Anglos)." The claim is that their parenthood is a scam, something they are doing to put one over on deserving citizens.

The legal and social status of pregnant women has of course grown thornier since the Supreme Court's anti-abortion *Dobbs* decision.

I am thankful that my experience of becoming a mother was the experience of making a deeply personal choice. Choosing parenthood felt like a vindication of my values and my relationship with my spouse and other private, precious things. All of that is still true about the reproductive and family choices I made, I think. However, the ability of a Republican government and a relentlessly strategic conservative legal movement—and here I am talking specifically about those in power, mostly men, mostly white—to make pregnancy and birth mandatory, casts a harsher retrospective light on those feelings of agency. Choosing to bear children felt like this key moment of autonomy and citizenship and adult actualization—but maybe I was never exercising a right at all, and merely being pacified with a revocable privilege.

And there's of course this other echo of the sucker dynamic at play where women are asking for abortion rights, laying out the rights and harms at stake, and being met with skepticism, like they are high school kids trying to get out of doing chores. Women are frantically making the case: pregnancy will be harmful to my health; this pregnancy is too risky; any pregnancy is a big deal! And the response, individually and structurally, is implicitly: *Come on, how hard could it really be? How much danger are you really in? Can you prove it?* There's this subterranean accusation in abortion debates where women seem to be suspected of trying to have sex for free—which of course is only a thing if what you really think is that women should pay for having sex.

When the *Dobbs* opinion was released in its final form, I was sad and scared, yes, but also surprised to find myself feeling personally insulted. It was not just the blatant hoarding of patriarchal power, the crystal-clear knowledge that our society will make women do things it would never dream of imposing on men (though, that too!). It was something about the glib, sloppy opinion—did they just cite to Sir Matthew Hale? The witch trial guy???—that had a sinister wink to it. Like, *I don't know what you thought, but a right that's this easy to take away was never yours to begin with.*

The Backups

Whether they are respected, or scrutinized, or derogated, or patronized, as a class, mothers are very reliable. They are more likely than fathers to be custodial parents; they spend more time parenting even after accounting for paid labor force participation; they allocate more of their pay to raising children. When a child is ill or disabled, or when a pandemic strikes, it is mothers who take time off from work. This is true for women across income levels and across demographic categories. Mothers show up.

During the early months of the COVID-19 pandemic, the work and family correspondent for the *New York Times*, Claire Cain Miller, wrote an article titled "When Schools Closed, Americans Turned to Their Usual Backup Plans: Mothers." Countless examples offer a portrait of a country that relies on its mothers as suckers (or saints!) of last resort.

Up until the late 1970s, the United Kingdom had a universal family benefit called the child tax allowance. This was a tax credit for families with children, and it was distributed by way of a reduction in withholdings from the paychecks of married fathers. In 1979, the child tax allowance was replaced with the child benefit, which was distributed in the form of a nontaxable weekly payment to the mother. One Member of Parliament argued against the change on the grounds that it would "take money out of the husband's pocket on the Friday and put it into the wife's purse on the following Tuesday. Far from being a child benefit scheme, it looks like a father dis-benefit scheme."

Shelly Lundberg, Robert Pollak, and Terence Wales are research scholars who study the economics of family life. When they looked at the policy change in the U.K., they saw an opportunity to test an old economic theory with data about family spending patterns before and after the new law. The theory that the researchers challenged was called the "common preference model," and it essentially said that it doesn't matter who in a household earns money, because the family budget is pooled. (This was the kind of argument that people would use for why it was fine to inflate men's salaries and keep women's wages low: If they live together, who cares how they get paid as long as the total is high enough?) "In contrast," the economists speculated politely, "individual utility models of the household permit the income received or controlled by one family member (for example, the wife) to have a different effect on consumption and time allocation than income received by another (for example, the husband)."

This circumspect description had a backstory. They were arguing that giving money to the father might yield different family spending than giving money to the mother. The original child tax allowance had been widely viewed as a unintentionally personal bonus for fathers—one that went directly from the Friday paycheck to the local pub—not to the children at all.

Lundberg, Pollak, and Wales predicted that paying more money to mothers, without changing the total family income, would benefit children overall. To test their prediction, they used data from the U.K. Family Expenditure Survey, which had tracked the weekly spending patterns for a large sample of British families for two periods: 1973 to 1976 and 1980 to 1990. They saw that there was no use looking at expenditures on items like books and food, because even if a family started spending more money on food, you wouldn't know whether that benefit accrued to the kids specifically; it could just be mothers were buying more bonbons for themselves.

However, the money spent on clothing was specified by category in the data set: men, women, and children. The economists decided to look at the ratio of weekly expenditures on children's clothing to weekly expenditures on men's clothing, comparing the pre-benefit to the post-benefit era. (They used a ratio because that way they could see something like the proportion of the budget spent for children, rather than the total budget, which itself might go up or down because of various factors like family income or inflation.)

For families with two children or more, the ratio changed dramatically. In the early period, a family of four would spend 26 percent more on children's clothing than men's clothing. Once the benefit went to the mother, though, the family spent 63 percent more on the kids than on the dad.

Vous travaillez pour l'armée, madame. Thank you for your service, ladies.

❖

Even as I write out this sort of can't-win narrative, perhaps I don't mean it. I wanted to be a mother, and I am, and now I have children who are my most profound pride. Surely I have won? The choices I make as a parent, as a mother, are intentional, and at some serious and deep level of my self, I am vain enough to think of them as also moral. I don't think I have just internalized benevolent sexism—women are pure, women are ethical—but I know that I am party to a set of parenting deals that I believe I should take, that I know I will take again, that I pride myself in taking. And yet I know that something about the choice set—constrained, skewed—is unfair.

The latter epiphany did not really gel for me until my sister had her first child. My sister is four years my junior, and possesses a litany of qualities I do not—athleticism, patience, a certain stoicism, not to mention the small motor skills of an actual surgeon. But sometimes she feels like another self. We look and sound alike, both tall and prone to mumbling. When she is slighted, I take it personally. She is a gynecologic oncologist, and she had been preparing and working for women's health for years, almost decades, before she had her son—studying for exams every vacation we've ever taken, working endless nights delivering babies in residency, slogging through a grueling fellowship a thousand miles from family, driving to the hospital in the cold dark of rural New England for emergency surgeries. She is an objectively impressive person whom I admire deeply. She takes responsibility for people and families in pain. Sometimes she operates and prescribes chemotherapy and sends them home, healthy, back into their lives.

Other times she gives them terrible news and bears witness to the end of another woman's life. For me this would be life changing if I had to do it once, but she does it all the time. She does these things and yet she is still genial, generous, present.

When the pandemic hit, she was at the end of her first trimester. Her hospital warned physicians that there was not enough personal protective equipment for everyday use. My mother sewed cloth masks in bulk and mailed them from afar. The pandemic wore on as she switched to larger scrubs—not maternity scrubs, as I assumed, but just bigger sizes, because apparently her hospital did not stock the styles designed for pregnant doctors and nurses. (Maternity scrubs can be special-ordered, but like many surgeons she goes to work in street clothes and changes into hospital-issued scrubs, that are laundered and disinfected on-site, right before she operates.) She tried to explain to me the very specific challenges of the billowing nonmaternity scrubs on a pregnant body; the physics of the whole operation meant that you had to wear suspenders if you wanted to be able to keep anything heavy, like a phone, in the pants pockets. She was funny and matter-of-fact, and she never extracted a price.

I wanted her patients, and her coworkers—every hospital administrator and government official and passing acquaintance—to shower her with gratitude and beg her forgiveness all at once. I wanted them to fall at her feet. She took call; she did surgeries that kept her on her feet for hours; she regulated her breathing as the mask capped her airflow and the baby pushed on her diaphragm. She had a gorgeous little boy and returned to work, at the height of the second wave, with a breast pump in her lunch bag.

There is a sucker's puzzle at the heart of mothering. Breast is actually best. Diverting the family budget to care for children is right. Being a mother is *good work*. In my personal moral universe, for sure, this job is what everything else revolves around. And/ but: once you know this about me, I am a perfect mark. Fool me once, twice, three times—who's counting?

CHAPTER 9

The Sucker and the Self

The first year of the pandemic, my fourth grader and I did a lot of walking around Philadelphia. This was on my insistence, for "fresh air" or "stretching our legs." I tried to keep her occupied with conversation. Once I let her describe to me, room by room, the spatial layout of the online game Among Us—"Now you go down and you're in the Med Bay. To the left there's a vent. Then you go out to the right and you're in Electrical. Okay, now you can walk down a hall. . . . And there's another vent!"—which got us, like, eight long blocks.

To keep her occupied during one of these walks, I told her she could be a subject in psychology research, and I started with the classic: the good old Ultimatum Game. I narrated it with maximum dramatic effect. "You're paired up. Your partner has ten dollars to share. You look at the message to see what you got, and you can't believe it: They're only giving you ONE DOLLAR? What do you do???" I tried to sound anguished.

She was nonchalant. "Well, a dollar is better than no dollars.

Keep the dollar," she responded easily. I protested: What about fairness? What about honor?! Revenge!!!

"What? Mom? No." She regarded me curiously. "What do you care if they get more than you?"

It's a pretty good question! What do I care indeed?

When she was much younger, still in nursery school, we used to joke that she was our little utilitarian. And she really wanted to know what it would get me, how it would make my life better, to decline that dollar. What do I care if I'm playing the fool? What are the stakes? How do I account for the pain of feeling duped, or the comforts of self-protection? The answers to these questions are important, because they excavate truths about our moral selves. Who do I want to be? What are my obligations, and to whom? How we figure out what our goals are, and how the sucker fear gets in the way, is at the heart of moral reasoning.

Rational Fools

When I talk to people about the fear of being a sucker, they often want to talk about rationality. Isn't it *rational* to avoid getting duped? Am I saying that the fear is *irrational*? As an academic, I am professionally obligated to respond, "Well, it depends," and unfortunately in this case I mean it. A fear is more or less rational, depending on what you're trying to do. For thinking clearly through sucker problems, you have to be explicit about the goals.

If I am a player in a Public Goods Game, for example, or a Prisoner's Dilemma, I can't evaluate my own strategy without knowing what I want out of the game. What values am I trying to vindicate with my choice? Game theory starts with the idea of a rational actor. As a shorthand, economists assume that most people are self-interested maximizers, meaning that, all else being equal, people want to make the choice that will maximize their own welfare. This often means they are trying to get as much money as possible, but the idea of rationality, even in economics, is more capacious.

It just refers loosely to the idea of having goals and choosing behaviors in line with those goals. The goals can be whatever goals you want. Maybe you don't want to make money; you want to distribute money evenly. Maybe you don't want to make money; you want to make friends. I don't have a theory of whether it is rational or irrational to be fearful of playing the fool; it just depends on the situation.

The fear of playing the sucker can make it harder to read your own moral compass, muddying the picture to make it seem like housing for the homeless is a trap, or cooperating with others is a weakness. In example after example in this book, from welfare benefits to racial violence to simple lab games, I have puzzled over situations in which people appear to be getting in their own way. When my mother's parents were in their precipitous decline, one of the most confounding facts of their paranoia was that they were rejecting the exact people whom they needed the most. Their deep suspicion made it harder to do what they most wanted, which was to keep living in their own home for as long as possible. For them and for countless others, the sugrophobic whisper was a constraint on personal and moral agency.

In an Ultimatum Game, for example, if your goal is to make money, then the rational choice is to take a small offer. Refusing the low offer is, at best, a Pyrrhic victory. You didn't play the fool, but that's the only prize; you made yourself and someone else worse off. If your goal is to be a kind person, the rational thing to do in the Ralph's Garage situation from the beginning of the book is to call the number and report the accident.

When I was taking graduate courses in psychology, we covered this obscure game called the Minimal Effort Game, a quietly diabolical setup that used the sucker fear like a straitjacket. What made it unusual is that the payout structure stacked the deck so that every possible value was on one side of the ledger—except fear.

The rules were simple but the incentives were confounding, although anyone who has ever been on a team with a weak link will quickly grasp the setup.

Imagine you are part of a group project, in this case for a grade. The project has seven group members and each member is taking one section of the presentation. Each section will be graded individually, but the final grade for each member of the group will be the grade assigned to the *worst section* of the presentation. So if there are six As and one C, everyone gets a C. The team's outcome depends entirely on the weakest link.

This is not a Tragedy of the Commons setup, where a selfish person might "win" by free riding on everyone else's cooperation. The Tragedy of the Commons would be more like the standard group project I talked about in chapter 1, where the grade is the average of everyone's performance, so a person could do his part poorly but still walk away with an A-minus. In the weak-link scheme, however, the person who shirks still gets the C, though they really prefer the A. But compared to their colleagues, they are a little better off. Some people put in A-level effort and get a C, while the weak link gets his C for free.

Suddenly the sucker stakes are clear. As a player, you know for sure that an A with hard work is your first choice: you care a lot about that grade and you don't even mind the work. A C with no effort will feel terrible. Still, if you work hard and get an A for your part, you are at risk of getting a C on the project overall even so. Then you're the try-hard who put in lots of work but nonetheless got a bad grade. You get the C *and* play the fool.

The Minimal Effort Game was run with seven players, and each player was asked to choose an "effort level" from 1 to 7. Each effort level was worth 10 cents, and participants would be paid according to their total points over a series of rounds.

The perversity of this arrangement was laid out in a payout matrix. Each player's reward was based on two questions: What did the least-contributing player do, and what did you do? A player got docked points for every effort level they contributed above the weakest link.

The clear best thing for everyone to do was to choose 7. Choosing 7 has no costs to any player, and if everyone chooses 7, every-

	Minimum Choice in Group							
		1	2	3	4	5	6	7
Own Choice	1	70						
	2	60	80					
	3	50	70	90				
	4	40	60	80	100			
	5	30	50	70	90	110		
	6	20	40	60	80	100	120	
	7	10	30	50	70	90	110	130

one gets 130 points, which is the most possible points. There's no way to "win" by being selfish; the only way to get big points is to go for 7. It confounds any attempt at a rational explanation for not cooperating. Want to make money yourself? Cooperate. Want to maximize the public good? Cooperate. Want to be generous to others? Cooperate.

But if you look at the table, you know that in any given round the person who cooperates the hardest ends up looking like a chump. So imagine for a minute that I choose 6. Two people choose 5 and one person chooses 4. You can see what happens by placing it within the table: Own Choice is 6 and Minimum Choice is 4. I get 80 and everyone else in my group gets 90 or more points.

So what actually happens? In the lab, people started out at various levels across the board, playing multiple rounds and adjusting their strategy over time. In the end, they almost always converged at 1. The fear of playing the sucker led inexorably to a case of Everybody Loses.

There are two kinds of "mistakes" that a player could make in this game, but the players only seemed to learn from one of them. People who cooperated and got burned adjusted downward, but people who defected and missed out on the big prize didn't adjust upward. In a sucker's mess, the players over time brought all their focus to the comparative payoffs rather than the absolute payoffs: they were so afraid of being the loser that they stopped trying to win at all, much less cooperate. Even without any real costs to co-operation, the relentlessness of the sucker setup was fatal to the group success.

From a psychological perspective, the core challenge of sucker's dilemmas is that they are hard problems. You're trying to do the right thing by some metric, but doing the right thing requires complex engagement with social perception, risk evaluation, and emotional processes. The Minimal Effort Game, like the other games, is a little world unto itself. The sucker who plays the high number only to take the lowest payout is on the outs; he's the group stooge.

But then I think about my daughter: What do you care if you're the sucker, at least in that situation? It's a narrow way of looking at the game, and it ruins it; players who are trying to "win" rather than "do well" wind up with an outcome that's bad by any measure. The group does badly *and* the individuals do badly, at least compared to the readily available outcome they universally want (130 For All!).

The Minimal Effort Game is a hard problem. It is cognitively demanding: just figuring out that payout table takes concentration. And it's also unusual, in the sense that relative success and absolute success are positively correlated in most familiar situations. There is a saying in the legal world that "hard cases make bad law." When the facts are really weird and there are many competing values, the judge can solve the dispute case in front of her (maybe), but the resulting precedent just doesn't analogize well to more typical scenarios, the kinds of scenarios we are most concerned with making law about.

In the field of judgment and decision-making, this same principle applies. When reasoning tasks are complicated, we are both more reliant on our shortcuts and assumptions for simplification, but they are paradoxically less relevant, having been formed to deal with the rule rather than the exception.

These shortcuts, often called "heuristics," have a recognizable pattern: they take a hard question and swap it with something related but much easier. The brain does this automatically, outside of conscious awareness.

For example, imagine that I give you an example of some kind of event and ask you to estimate its frequency, something like "How often does COVID-19 result in hospitalization?" There are epidemiological resources that will give you the answer to that question, not to mention Google, but if you're just reasoning the question on your own, it's very hard. Most people will do a little mental switch; when it's challenging to answer "How common is this situation?" they swap in something else: "How easy is it for me to think of examples of this situation?" So "How frequent is this?" becomes "How mentally available are examples of this?"

This particular swap is a famous phenomenon called the "availability heuristic." For many questions, the mental availability for statistical-frequency substitution is very helpful. If I'm asking, "How often does my daughter come home from school with her lunch uneaten?" I can search my mind for recent memories of opening up her lunch box, finding a soggy old sandwich, and throwing it out. If I can't remember the last time this happened, or have to really think back, I have a pretty good sense: this is a rare event. If I can recall three instances right off the top of my head, it's a good bet this is a common occurrence.

But for some questions the heuristic leads us predictably wrong. It explains, for example, why people tend to overestimate the frequency of really shocking events: shocking events are always on the news, people talk about them, and they have such emotional resonance that they create vivid memories. (Most of us overestimate the frequency of plane crashes, for example, and most of us underestimate the frequency of pulmonary embolisms. It turns out that even doctors who encounter one patient with a rare disease are more likely to mistakenly diagnose that same disease soon after, because it's so hard even for experts to mentally disentangle frequency from cognitive availability.)

In unexpected places, you may be able to see yourself relying on a misfitted shortcut, including, I think, a status heuristic for sucker puzzles. At the start of this book, I raised the possibility of an affirmative goal, something like integrity. If you ask people about their goals in complex situations—economic, moral, and social dilemmas—that is where they will often land. What choice vindicates my integrity? But sometimes that question gets implicitly swapped out, unnoticed, for an easier query: What choice vindicates my status? What choice is going to feel better in the moment? As a heuristic, the replacement is structurally classic: integrity is complex and resource-intensive but status is typically intuitive and automatic. I might not know how to do good, but I know how to do better than you.

Good Job

Sometimes you have a job or a role, the job has parameters, and the goal is simply to execute. Showing up, doing the work, and getting the thing done with competence and efficiency is often what it means to do the "right" thing. In those situations, we want to get the sucker fear right-sized, or well calibrated, because it can work mischief with basic efficacy. Overinvesting in sugrophobia can get in the way; it distracts from the real goal.

This straightforward kind of mistake exasperated critics of the Trump administration. Trump's obsession with the United States being "laughed at," stoked in his Manhattan real estate days, was not just irrelevant in the international trade context but actively destructive to his own purported interests in protecting American consumers. In 2018, President Trump announced a new set of trade tariffs. His reasoning? The United States was being "ripped off" by China: "Other countries have become very spoiled because they always got 100 percent of whatever they wanted from the United States," he said, "but we can't allow that to happen anymore." He declared trade wars good, predicted they would be easy to win, and then promptly lost them, costing American consumers billions of dollars. Experts on trade viewed his position as wildly

wrongheaded: international trade is a complex system of interde-
pendencies that leverage the comparative advantages of different
national economies; it's not a schoolyard fight. Trump would have
made himself and his constituents better off by cooperating.

In fact, sometimes a willingness to risk playing the fool, or look
like a sucker, can be a useful tool for dispute resolution. I joke
sometimes that my leadership persona is "affable fool," but I sort
of mean it. When I was an associate dean, I had no real authority,
and mostly just a lot of low-level embroilment in faculty nonsense.
I did have goals, though. I had hopes for the hiring and retention
of exciting scholars and teachers, and I felt that I knew what student services and resources we needed. A few months into the job,
I decided to just lean into the fact that I do not care very much if
people I work with think I'm a sucker, at least in the short term.
There are contexts in which the price of looking foolish is too high,
but this job wasn't one of them for me. Sometimes if I write that
memo (not my job) or draft that schedule (not my job), it's the
fastest way to the proposal or the itinerary or whatever the thing
is that I actually want.

Experimentally, there is evidence that when some people are
willing to risk weakness, their example can set a norm that helps
the group commit to cooperation and achieve its goals. Mark J.
Weber and J. Keith Murnighan, both business school professors
and behavioral decision researchers, wanted to see what happens
when some players consistently behave cooperatively, even if
others are selfish. They conceived of their study as a type of experimental examination of the famous Margaret Mead quotation
"Never doubt that a small group of thoughtful, committed citizens
can change the world. Indeed, it is the only thing that ever has."
They called their paper "Suckers or Saviors?"

They asked players to participate in a standard Public Goods
Game; each of four players would get money and the chance to
contribute it to the communal pot. The amount in the pot would be
multiplied and evenly redistributed as usual.

There were a couple of twists to their method, though. The first was that the players would participate in not just one round of the game but twenty. Other studies that had used multiple rounds had routinely found that playing the game repeatedly would result in less cooperation with each iteration, and the researchers wanted to see if they could reverse that trend. To do that, they assigned every foursome a plant: one confederate who would contribute to the group pot no matter what.

For people trying to figure out whether to go for the cooperative choices—a choice they may find appealing, but risky on the sucker front—the idea was that a dogged, reliable contributor could make the risk feel less acute. It's not so bad to mistakenly cooperate when someone else is playing that same strategy; at least you're not a pariah. This proved true: including one consistently cheerful sucker affected the choices of everyone else in the study. People who played with a frequent contributor were more likely to contribute themselves, and games that included the consistent contributor did not devolve the way that multi-round Public Goods Games normally would. The sucker was the savior.

Sometimes you even have to play the sucker to save yourself. Like most people my age, I have intermittently searched for better health in one domain or another: pain relief, fertility, energy, good sleep, good skin, and so on. Many of these episodes are just embarrassing or personal enough that I'm not going to describe them in a book like this—sorry, or you're welcome, according to your tastes—but in certain respects, incident to incident, they shared a common sucker's form. Each promising new pill, device, or diet would bring fresh optimism and then the pall of suspicion. Is this a scam? Am I the mark?

A few years ago, amid a confusing cascade of health issues, I had a prolonged bout of terrible joint pain. I tried some normal things and wasn't a fit for some of the other normal things. I tried to ignore the pain and felt humiliated when I could not. I was advised to try meditating and avoiding stress. Thus left to my own devices, I searched online, sometimes waking in the predawn, searching for unlikely online answers while my ankles and wrists throbbed chaotically. A week after one of these appointments with Dr. Google, I walked out of work and drove to an office park in New Jersey and then watched a nurse make sixteen careful injections of bee venom into my arms and legs.

The story was not complicated. I had come across an article about bee venom and inflammation. I think I read someone's first person account, and it made enough sense to me, or at least it sounded earnest and hopeful. I found a practitioner who offered a variety of unusual treatments, including the bees. The office seemed clean and professional. I gave it a try. I left the office with a row of nickel-sized red welts along each forearm, visible evidence that I had permitted—nay, requested—a series of purportedly therapeutic beestings.

I don't know if the bee venom helped. I experienced mild relief after each session of injections. At the very least, it was distracting. It sort of felt like the stings would reroute my body's misguided immunologic response away from my tender joints and over to the low-stakes sting sites. Overall, the pain did in fact subside, roughly contemporaneous with the beesting sessions, but correlation/causation/blah blah blah . . .

I told very few people about my adventurously empirical approach to pain relief. It's a confusing story and mostly embarrassing. If I told you how much it cost, you would almost certainly think worse of me. If I am being one step more honest I will tell you that there were crystals involved.

But I will tell you something else, which is that I was not in any material sense duped. I knew exactly what I was signing up for: an expensive long shot. The whole enterprise may have been nonsense, but that was part of my explicit risk assessment. The pain in my body had begun to interfere, in significant ways, with my ability to concentrate at work, to be present with my kids, to derive pleasure from my life. (I have described this as a personal rather than a moral dilemma, but I am not even sure this is right. As more serious philosophers and activists have argued, there are ethical implications to one's own health and well-being, perhaps especially for caregivers.) The available remedies were either not working (ibuprofen, Tylenol) or had an unacceptable side effect profile (opioids). The same way that you might buy a low-odds raffle ticket for a big prize, I knew a 5 percent chance of improvement—or even a 1 percent chance—was worth the cost, at least to me.

I think most people have had some sort of experience like this, though often for lower-stakes problems and more plausible interventions. My husband, a lifelong insomniac, read a book called *The Honey Diet* when the kids were little, and for the next year insisted on a honey-vinegar concoction before bed each night. A few years ago, friends feeling the creep of middle-age fatigue started putting butter in their morning coffee, convinced by the Bulletproof empire. The mild recriminations are also familiar. "I see you've bought into the hype," you might hear. "What's the newest snake oil cure-all? Haha." If you're the one whose turmeric supplement or CBD lotion is the target of contempt, the insult feels sharp: what a sucker.

Every once in a while during my beesting journey, someone I'd confide in would say something about placebo effects, the most medicalized (and misunderstood) sucker's mistake. Your whole body bought the con! But here's the thing: even though I got the derisive drift, I also found the accusation illogical. Why *wouldn't* I want a placebo effect? A placebo sounds delightful. If my brain is willing to block its own pain processes, I don't care if the relief I feel is the result of a direct biochemical shift in response to a treatment, or "just" a placebo. What difference does it make? You only get one body, and mine was making me miserable. Please, fool me, I'm begging you.

Better Living Through Accounting

Playing the fool is not rational or irrational; it's just one more consideration among others. One way to decide how it factors into a choice is to name it and give it a line on the ledger. Sometimes all it takes is an explicit reckoning with a hard decision: What are my goals and what are my options? When that cost-benefit analysis happens automatically, the sucker fear takes a lot of space; as you may have noticed, the whole point of this book is that it's an unusually salient fear. But salience isn't set in stone, and attention is flexible. The way to de-weaponize can be straightforward: Take inventory. Run the numbers. Account.

In chapter 2, I described an example of weaponization in a common work or school situation. Someone asks for special accommodations to deal with a personal emergency, like a student who requests an extension on a final paper because a family member has died.*

With the student in my office, making her request, my mind is racing to aggregate a bunch of judgments. I am tallying my answers to these questions: Does this situation warrant an extension, is she telling the truth, and what are the administrative costs of granting or refusing the extension? My automatic processing systems are going to be challenged by this task, and my mind will orient itself toward whatever factors are most salient. The lurking possibility that I might be duped does a lot of undercover work in my reasoning, whether I notice it or not, and with the right trigger points being pressed, I might be inclined to be rigid about the deadline for fear that I might get hustled.

However, if my goal is to be good at being a professor—and it is—this is not a good outcome. I have a job with parameters. I am trying to teach students contract law, yes, but also to mentor, to welcome, to model something professional and sensible and humane. My real fears about mistakes I might make as a teacher are things like not preparing them well for the bar exam, shortchanging them in terms of my attention to their progress, or failing to treat them with respect. Preventing students from getting undeserved extensions is something I would care about a bit but largely

---

* I should be explicit here that as a matter of my actual life as a teacher, there is no situation in which a student would tell me they had a death in the family and I would respond with anything other than total credulity and accommodation.

at the margins. As a minor value, it shouldn't get to occupy a lot of space in my decision. But if I'm not careful, it will; it's hard to ignore that lurking fear, that hint of dread.

There is an intervention from cognitive psychology that counsels something that seems too mechanical to work: Run the numbers. This decision-making approach, called multi-attribute utility theory (MAUT), is both more profound and less complex than it sounds. The idea is that we are often trying to make decisions where multiple goals or values are at stake. If I am choosing whom to hire for a job I've posted, for example, I might want to choose someone who is kind to others as well as someone who is efficient. If I am deciding how or whether to accommodate a distressed student, I might think of myself as having two goals: compassion and cheating prevention.

Each choice in front of me—grant the extension or resist it—serves my goals to a greater or lesser extent, and I need a way to compare the options so that I can see which one serves my goals best, all things considered. The insight of MAUT is that I can be my own decision algorithm. I can create a rough cost-benefit calculation with a table, and even assign numbers, and they really help.

Here, I have two choices: conditional extension (Can I get some paperwork?) and immediate extension (I'm so sorry; let's move the deadline back two weeks). I have two values: compassion and cheat prevention. Let's say for each choice I score it on a 1-to-100 scale. How well does this choice achieve this goal? I come up with the following values:

	Compassion	Cheat Prevention
Conditional Extension	30	100
Immediate Extension	100	0

Now ultimately the idea is going to be that the choice with the highest score is the one I am going to pick, and it looks like conditional extension is winning. It fully achieves cheat prevention and partially achieves compassion, whereas immediate extension fully achieves one goal and totally fails the other.

Mapped out like this, it is clear to me that my accounting is wrong; it doesn't reflect my actual utilities. The fear that I would play the sucker is a much less serious and important value than the duty to treat students with dignity and humanity. But the other core insight of MAUT is that when values themselves are not equal, I can account for the differences in my scoring.

I would much rather be mistakenly generous than mistakenly skeptical. Preventing cheating matters a little bit to me, but honestly very little, at least in the context of a school with plenty of other mechanisms for enforcing academic integrity (honor codes, blind grading, and more). It's not nothing; I don't want students to walk all over me. But to do the right thing, given my conception of my role, I need to re-weight my answers.

So I decide: compassion is four times more important to me than cheating prevention in this situation. I change my chart accordingly, reducing the values in the cheat prevention column by literally just multiplying them each by a quarter.

	Compassion (100 percent)	Prevent Cheating (25 percent)
No Extension	30	25
Extension	100	0

This makes it much easier to see what I want to do. The cheat prevention column reflects my fear of playing the sucker and locates it within my actual ethical code.

I have used tables like this for a whole range of personal decisions: what house should we buy, what day care should we choose, whom should we hire for this job? It can feel artificial, even silly, to assign number weights to my own values. We don't *know* if the student is cheating or what the right answer is, or even the right probability to affix on different outcomes. And is it really *true* that in some meaningful way I value fraud detection at 25 percent of student compassion? Why not half, or a tenth? The truth is I don't know with any certainty. But the exercise improves my decision-making, even if it's based on imperfect measures.

The reason that a little fear like sugrophobia gets to take up a lot of space in decision-making is because it normally doesn't get accounted for at all; it just seeps in like a contaminant. Disaggregated decision-making like the MAUT table forces it back into its own cell, cabins its effects from the rest of the values and utilities.

This phenomenon is part of a body of literature in psychology that compares holistic decision-making—what's your gut tell you?—to disaggregated analysis. The major insight is that people are good at evaluating attribute by attribute, but terrible at combining those evaluations into a whole. Robyn Dawes, one of the original investigators on the Public Goods experiments, wrote a paper in 1979 with the somewhat psychedelic title "The Robust Beauty of Improper Linear Models in Decision Making," in which he recounted the benefits of even the most amateurish pros-and-cons listing over holistic judgment, a method he called "bootstrapping." By listing my judgments one at a time and then adding them up, I can pull my own reasoning up by its bootstraps, so to speak, or improve it without actually relying on any external resources or support. People make better decisions when they judge each value intuitively but add those constitutive judgments together mechanically.

If I am deciding how to treat a student, or whether to donate cash or food to a food bank, or even just whether to take a lowball offer in an Ultimatum Game, I have reason to be worried: Am I the sucker here? And I think at some level that the fear is something I do take seriously. The question I want to make sure to address is *how* seriously. I prefer not to play the fool, but compared to alienating a grieving student, that preference is trivial. Or, I prefer not to play the fool, but as long as I am giving money to alleviate hunger, my goal is to alleviate the maximum hunger per dollar spent. What is my goal and how do I get there?

Moral Agency

I have technocratic tendencies: I like cost-benefit analysis; I really think we can run values and information through our own reorganized minds and get something better than the mush we started with. The fear of playing the sucker is a distraction, or a distortion, that undermines efficacy in the day-to-day. But I think the challenge is deeper than that, at least for the moral self. The fear of playing the sucker is a constraint on the moral imagination. I want to do a good job, yes, but I also want to do the right job. I want the work I do to have integrity.

The example of the distressed student is instructive at a deeper level, or maybe two deeper levels. When I run that example through my own values, it is clear to me that the right response, when someone in my fiduciary orbit appears and says that they are grieving, is to react with immediate compassion. Let's imagine, though, that I have had run-ins with this student before. I have reason to think that she might exaggerate her situation, because I know she is behind and increasingly frantic. She has turned other work in late. The probability that my requirement of documentation would in fact successfully dissuade her from cheating in this way is higher than average. In that case, the MAUT math might be more equivocal—and I think that's misleading.

In a student-faculty transaction, I have a lot of power. I'm doing the grading; I'm writing the recommendations; I'm the one with tenure and institutional buy-in if there's a fight. If a student comes to me in distress, even if the story itself feels a little fishy, there is a core moral force to our interaction. I want to help her take her next step. There are many ways I can do this, but asking for documentation is not one of them. The student claiming bereavement may be telling me something false that she thinks I'll accept because the true source of her distress is more private or more shameful. She may have gotten herself into a bad position and can't figure out how to get out.

My goal isn't just efficient vindication of a set of professional priorities; my goal is moral integrity. Often if I'm worried about being played, I'm indulging in something self-absorbed or vain rather than moral per se. Sugrophobia is a challenging worry because it's such an attention hog, but it doesn't actually deserve much moral regard.

In some ways it is easier to see the moral hollowness in a counterexample. In 2020, the *Atlantic* published an article quoting Donald Trump's petulant complaints that fallen soldiers were fools.

In a conversation with senior staff members on the morning of the scheduled visit, Trump said, "Why should I go to that cemetery? It's filled with losers." In a separate conversation on that same trip, Trump referred to the more than 1,800 marines who lost their lives at Belleau Wood as "suckers" for getting killed.

Many dynamics can be slyly cast as sucker's games—but soldiers??? For most Americans, soldiers are the very model of personal integrity. Trump's moral framing explicitly and sincerely rejected personal sacrifice for a common good; as Jeffrey Goldberg wrote in that piece: "Trump finds the notion of military service difficult to understand, and the idea of volunteering to serve especially incomprehensible."

I think most people who hear this claim—fallen soldiers are losers—get that it is fundamentally mistaken. Whatever you think about the moral righteousness of World War I, or even armed conflict generally, the core of that transaction, between the individual soldier and the country, is about a deeply cooperative and prosocial sacrifice.

There is a doctrine in contracts called the duty of good faith and fair dealing. First-year law students are often surprised when they encounter it, because they expected contract law to be formal and rigid. The duty of good faith says something like: This agreement has integrity. It's not just every man for himself; the deal has a moral coherence. The doctrine operates as a repudiation of the caveat emptor story. For example, in the 1960s a man named Orville Fortune was employed, and then fired, by a company called National Cash Register. His job was to sell cash registers to businesses, and the terms of his employment were clear: he was an at-will employee, so he could be fired or laid off at any time. He made a modest base salary and then worked on commission, which meant he received a percentage when the deal was made, and then another portion when the registers were delivered, installed, and paid for. With large accounts, that process could take many months.

A decade into his tenure at NCR, Fortune had just landed a big sale, with a deal to provide cash registers for a chain of First National Bank locations. Almost immediately after he landed the deal, he was fired—losing not only the remainder of the commission on the sales, but also the very real prospect of ongoing sales with the bank in coming years. Under the terms of his employment contract, it didn't look good for him. Indeed, he looked like a dupe, working steadily and loyally without any guarantees for all those years. The court refused this interpretation. It said to NCR, in essence: we do not agree that you made an exploitative contract. You and Fortune agreed that you could fire him at any time, yes, but you also agreed—as all parties must—to treat one another in good faith. You can fire him at any time *except* when you are using that power to take back something implied in the contract. If the deal is to compensate by commission, you are implicitly promising that all else being equal, he gets to see his sales through.

There is an integrity to staring the sucker narrative down. It's so easy to feel contempt for Fortune, and that contempt can obscure the parallel facts that NCR was behaving opportunistically, engaging in the kind of sharp practice that deserves is own reckoning.

Every once in a while, I encounter something like this in my personal or professional life, and it really is hard to know what to do, because it feels embarrassing to announce that you've agreed to be a fool. A few years ago I was invited by some hiring committees to visit at other law schools. "Visiting" sounds more informal than it is; the invitations are actually for trial runs, where you go to a school for a semester or a year—usually months, not days—and teach courses to their students and interact with their faculty. At the end you go home and wait to hear if they liked you or not. It is flattering to get the professional attention and also challenging to execute the performance. You have to be very *on*—charming, engaged, witty—with the new school's faculty for weeks in a row and all the while the rest of your family either waits at home short-staffed (difficult), or bumbles along in a random city together (also difficult).

I agreed to three such visits, and got zero job offers. In the end, I really did feel like an idiot. The schools that do the recruiting are typically full of praise: your work is so good, you are a rising star, etc. But on the other end, the way it works is you either get an offer or you hear nothing. Rejection by ghosting. By the end of my last stint, which I sort of knew was a dud even as it was happening, I felt pretty humiliated in the normal way—the visits are public, so among other things everyone you work with knows you just got rejected—but also in a more meta way. Why was I such a sore loser? I agreed to this whole thing and now I couldn't handle the risk I had volunteered for?

A couple of years passed, and a few more people told me about their own thorny feelings, how demoralized they had felt after a hiring visit, how long it had taken them to regroup. I felt bad that I had pretended my visits were okay, when I knew they weren't. Publicly and privately I decided to be frank, rather than gracious, about what the disruptions had cost me, how the visits had made me feel. Whatever vulnerability I experienced announcing my various humiliations in meetings or email chains, it was mostly just self-indulgent. I'm not actually vulnerable, and I should speak up. I have tenure, so it's very hard to fire me. I don't think I ever even wanted a different job so much as I wanted to be able to say I'd had the option, a fact that reflects even more poorly on my character.

My own embarrassment was an effective cooler, a reason, when people asked, to say things like: Oh, no, actually it didn't work out, but I made some really great connections! Or: I think we're in a holding pattern for now. But when the mark gets cooled out, the operator is free to keep operating. I think it's important to be able to say: I am embarrassed that I fell for this, but I'm not the only one to blame here. I'm ashamed that I might be complicit, but this practice/system/situation is exploitative. One of the imperatives of privilege is being willing to be explicit about things that feel bad to admit.

Exposure Therapy

Like other fears, sugrophobia has a predictable method for inflating its own importance in its host human. The fear consists of a set of neural links, so that when it is triggered, it creates a cascade of thoughts, feelings, and behaviors.

In cognitive behavioral therapy, therapists typically identify the thing that is terrifying the patient—snakes, say, or angry people— and ask them to track the ways that this fear affects behavior.

One of the embarrassingly banal facts of cognitive behavioral therapy is that they give you worksheets to fill out, and they, like, work. You fill in the blanks on a series of questions. This experience made me feel _____. It triggered an automatic thought of _____. An alternative response would be _____. The idea is to literally rewire the brain, to make a new connection between the triggering situation and the alternative response rather than the automatic response.

Automatic responses are plagued by a series of common mistakes. I have one of these worksheets, and even the appended list of common distortions feels almost specific to the sucker's reaction. It cautions against "emotional reasoning" and gives as an example "I feel like an idiot so I must be one," and also warns of inappropriate labeling, like "instead of saying, 'I made a mistake,' telling yourself, 'I'm a loser.'"

These automatic thoughts and feelings link directly to predictable behavior; this is something most of us know intuitively from our regular lives. If I am afraid of snakes (I am), I might avoid certain kinds of travel that I'd otherwise be interested in, or refuse to go hiking alone. I don't watch *Snakes on a Plane*. Those videos you sometimes see on YouTube of a person who finds a snake coming up through their bathroom plumbing? I would never use that bathroom again.

The automatic thoughts of a sucker—*I'm an idiot! What was I thinking? This is so embarrassing!*—can create the kind of avoidance that leads to an unwillingness to engage honestly.

These kinds of avoidance techniques are natural but not inevitable. The way to deal with my fear efficiently is exposure, not avoidance—to be explicit about what the likely risks are and what I can take. If I can't make myself go into the garage, I should get someone else to come and check. Watch videos with snakes; touch them when I have the chance. Admit to myself that I have been duped, talk about it, let it sit out in the open. Is there another way to describe what happened? Is the fallout as bad as I had feared?

Sometimes being a sucker is truly harmful. It can have material or social consequences that constitute meaningful losses for the mark. But often the fact of being duped doesn't need to be a big deal. It exists as just a feeling, and that feeling need not have priority over what really matters.

Only Connect

When I teach contracts to first-year law students, I give a pep talk on the last day. I say: You are doing a great job, and your success is important to me. Every year I toy with the idea of talking about the thing that feels the most true, which is that the feeling I have when I am teaching feels like a form of love, in the sense of universal loving-kindness.

I give my own pep talk based in part on a pep talk inadvertently given to me when I was a student. In graduate school, I had to be a teaching assistant as a condition of my stipend. I was a TA for my dissertation advisor, Jon, in his undergraduate course on judgment and decision-making. He taught his large lecture in a windowless hall flanked by heavy double doors on both sides, the kind with crash bars. Students would periodically leave mid-class for a range of normal reasons and then return, and every departure and reentry was bizarrely disruptive. The door would squeak open, squeal back, and then make a cacophonous banging noise when it reconnected. One student left almost every day mid-class to get water or something, letting the door slam behind him each time. I made some kind of frustrated remark to Jon after class one day; I thought it was egregiously inconsiderate. I asked him about his approach to difficult students.

Jon is a very no-nonsense kind of person, smart and sort of dry and unsentimental. He was a fantastic advisor, in every sense of the word. I was surprised when he told me, more or less, that he doesn't make that kind of evaluation at all; his student philosophy is "unconditional positive regard."

In any case, now this is how I talk to my students about what I aspire to.

Unconditional positive regard is a psychotherapeutic philosophy developed by Carl Rogers, the most prominent clinical psychologist of the twentieth century. It is characterized by an attitude of caring and valuing another person irrespective of their behavior or choices—not accepting the behavior itself but accepting the person. It is a theory of helping people on their path to self-actualization.

Like Jon, and like me, my own father is a teacher, now retired. He taught third grade for almost thirty years, and he was really good at it. At the end of each year, he would hold a classroom awards ceremony, the Third Grade Oscars. He printed two awards for each kid—one serious, and one funny. A kid who got Best Science Experiment might also get the Bucking Bronco Award, a prize for the child most frequently thrown from his desk chair by the force of his own energy. The kid who read the most books might also be recognized as Loudest Kickball Player. I remember, as a kid myself, seeing him give out those awards (my sister and I were sometimes allowed to visit and "help" that day—i.e., filch candy from the prize bag); I knew how good it felt to those students, and their parents, to be that seen and that welcomed. Unconditional positive regard in action.

When I went back to reread Rogers recently, I was struck by his explicit repudiation of the sucker fear:

> The kind of caring that the client-centered therapist desires to achieve is a gullible caring, in which clients are accepted as they say they are, not with a lurking suspicion in the therapist's mind that they may, in fact, be otherwise. This attitude is not stupidity on the therapist's part; it is the kind of attitude that is most likely to lead to trust . . .

I may not take that stance with me to the grocery store or the mortgage broker, but when I am teaching, advising, collaborating, or, obviously, parenting, these ideas—gullible caring, universal positive regard—can be a counterweight, a glimpse of another way to think of myself as a sucker in the world.

Of course, most interactions are lower-stakes, and the goal is not always love or intimacy. But I do take seriously the possibility that many human relationships, though fleeting, are ultimately about a search for connection. Aspiring to unconditional positive regard is a way of being specific with myself about a goal that can feel too airy or nebulous to make it onto the matrix. If you do a guided meditation, like on an app, they sometimes ask questions at the end like "Can you bring an attitude of loving-kindness to the other parts of your life?" Can I? I hope so.

Conclusion

When I was very young, we had family friends, a couple, who lived in a little storybook hut in the woods a couple towns over. They were small and older and they seemed like characters from a folktale. In fact, they had moved to Maine from New York City and the wife was a developmental psychologist, which was perhaps why to me she seemed so magical: she knew from little kids. Her name was Dorothy, and she was my mother's professional mentor and then best friend until she died, suddenly, of pancreatic cancer. She and my mother were both practicing psychotherapists in rural Maine in the early 1980s, which presented some real challenges insofar as most of their client base were non-believers. Dorothy was stalwart. "The thing about psychology is that it doesn't matter whether you believe in it; it's still happening all the time," she would say to cheer up my mother.

The same could be said of the fool's game: it doesn't matter what you're doing about it, it's happening all the time. Reckoning with fear, or avoiding a scam, or fighting back, or backing down—we can deal, but we can't escape. We are swimming in a sea of quasi-grifts and maybe-hustles no matter what. Like others, I often find myself caught up before I really know what's happening.

This is true in deep global ways and little quotidian ones. On a given day, I might agree to drive my son to school rather than making him take public transit, because he says he has "too much stuff to carry." Upon later inspection, he has, at most, a slightly overstuffed backpack. I get to work and write a letter of recommendation for a student that highlights his good qualities and omits to report that his academic achievement in my course was below average. While researching, I surely agree, electronically, to bring any legal disputes I may have with Apple, Google, JSTOR, the *New York Times*, and Amazon as claims in arbitration rather than state or federal court. I do this without ever reading any of the contracts or knowing what they contain. Midday, if I present at an online conference on Zoom, I use the "touch up" feature so that I look less haggard. After classes are over, I go to a late-afternoon faculty workshop. It is not interesting and not convenient, but I thought that we had agreed as a faculty that workshop attendance was a core shared value. I find there are only eight of us present out of forty-five.

Living in a society is complicated. I really did need to download an article from JSTOR, and I definitely couldn't have read the terms and conditions first. My son's claims about his bag, much like my claims about my promising student, are made in a relational context. There are layers of meaning, and norms of inference, and niceties, and small mercies. Sometimes we let things slide and grant some leeway.

It's not that we don't understand these dynamics exist, but it can be hard to see individual choices in context, even for ourselves. Roseanna Sommers and Vanessa K. Bohns are social psychologists who write a lot about consent and social pressure. They recently published a set of studies that made me sit up and reconsider what I thought I knew about social influence. They hired a couple of research assistants and recruited Cornell undergraduates to come in for a study. When the subjects arrived at the lab, the research assistant would do one of two things. Some people would be asked to fill out a questionnaire, supposedly for researchers exploring the feasibility of a method for a future study. The questionnaire asked:

Imagine that you were seated in a psychology lab, similar to the one you are in now, and an experimenter came in and said to you: "Before we begin the study, can you please unlock your phone and hand it to me? I'll just need to take your phone outside of the room for a moment to check for some things."

The subjects who answered that questionnaire were in what was called the forecasting condition. The other subjects were assigned to the experiencing condition. The Experiencers had that same research assistant greet them and then actually ask directly, "Can you please unlock your phone and hand it to me?" Subjects who asked follow-up questions ("Uh, why?") would be told, "I have a list of illegal apps I'd like to check for."

It's helpful to pause for a moment and imagine yourself unlocking your phone and handing it to a stranger to scroll through. I am not that private of a person, but I would *hate* to let someone look at my stuff. So if you ask me: Would you hand over your phone for a stranger to paw through it? my strong instinctual response would be: Absolutely Not. In this respect, I am very much like the Forecasters in the study; 73 percent of them said they would refuse.

But when the research assistant looked the subjects in the eye and just asked them straight-out to unlock their phones and hand them over, the talk proved cheap: 97 percent of the subjects handed their phones over to a stranger and let him take it outside the room. (The research assistant did not actually look through people's phones; he just counted to five outside and then came back in.)

What's wild about this study is the magnitude of the effect. It's not that compliance was a little bit higher than forecasted. It's that almost every subject complied and almost none of them thought they would. Three in four subjects expected to refuse; one in thirty actually did.

The thing about being afraid of being a fool is that you are definitely going to be a fool, at least some of the time, by some definition. But it is hard to see from the outside exactly what would be so difficult about refusing to unlock your phone for a stranger. It seems really easy to say no from afar, and that perception can be misleading. If you see me blithely clicking "I agree" without ever reading Amazon's consumer contract, you might think to yourself, *She needs to get it together.* But the system is set up to guide behavior, and Amazon is definitely telling me, Don't bother.

This has implications at the personal level, but it can really reverberate at the legal and political level. The issue comes up all the time in contracts. Someone signs a contract while they are buying a computer, a line of people behind them waiting impatiently. Later, it turns out the fourth page had a clause about a warranty limitation, and when it goes to court, a judge says that the customer had an opportunity to read it: they could have stepped out of the line or downloaded the fine print before they went to the store.

There is a frustrating case that I (like many others) use to teach this conundrum: it's a dispute between a Washington, D.C., furniture store and one of their customers from the early 1960s. The store, Walker-Thomas Furniture Company, would send its salesmen out to target poor neighborhoods, where they could expect to find residents without enough cash to buy furniture outright or enough social capital to access traditional forms of credit. The salesmen would bring along their installment contracts, which they would thrust out for the customers. The contract itself had a large, bold header: READ CONTRACT BEFORE SIGNING. But before the salesmen handed over the contract to customers, they would fold over the top so that all that remained visible was the signature line at the bottom. "Just sign," they would advise.

One of their customers was named Ora Lee Williams, and she had bought about $1,400 worth of furniture over a five-year period. She was down to owing only $164 when, in April of 1962, she purchased an expensive stereo and then started missing payments. Walker-Thomas sent agents to repossess the stereo—along with *all of the other furniture* that she had purchased from them since

1957. It turned out the contract that she had signed, unread, included a dense, lengthy clause that ended with the instruction that all payments "shall be credited pro rata on all outstanding leases, bills and accounts." This meant that every payment that Williams made over the years was spread across all of her purchases. Now, keep in mind that any furniture on which she owed a balance was subject to repossession if she defaulted on her payments, and the pro rata system meant that it was impossible to pay off any one item until every item was fully paid. So even if she had paid in full an amount that would easily cover everything in, say, her dining room and living room, every piece was subject to repossession once she started to default on the stereo payments. She sued. She claimed that the contract was unconscionable, that it should be voided, and she should get to keep the furniture she had paid for.

In court she was questioned by the attorney for the furniture company. Didn't you read the contract? they asked. How could you make all these purchases without reading what you were getting into?

Mrs. Williams was frustrated. "You are asking me about reading things I never had to read," she finally exclaimed.

She was right. Why was all this talk about reading the contract coming up now, when she knew and they knew that she had been explicitly discouraged from reading when it mattered? The clause had been written to be as opaque as possible, and then it had been literally opaque, covered with other sheets of paper. The court agreed: what Walker-Thomas had done was predatory, whether Williams had manifested assent or not.

Some kinds of scams are the cost of doing business—not that you have to accept them, but that you can't really avoid them. You can't read the news without agreeing to the fine print, and if you're Mrs. Williams, you can't get the furniture without signing an outrageous cross-collateralization clause. More broadly, you can't always remember to say no when your instinct is to be polite. You really can't avoid the snake in the grass forever; the idiom is precisely about a risk that's hard to detect. Wooden nickels are not just the cost of doing business; they're the reality of being human.

❖

In so many places in our lives, the fear of being a fool is covert and sublimated. We are carefully skirting it without ever acknowledging that it even exists, rearranging our selves and our society around a threat that doesn't even get a name. But there is an arena where it gets to take center stage, a star rather than a stagehand; that arena is love. Maybe people don't *want* to be suckers in love, but they know the risk is endemic to the enterprise—and the enterprise is existential. In love, they'll let uncertainty lie, admit and maybe forgive betrayals, and reckon candidly with the bounds of their own vulnerability.

Love songs are full of fools. Aretha Franklin alone joined a chain of fools, was "a fool for you" and threatened that her lover was "runnin' out of fools." Elvis was a fool, rushing in. Bill Withers was a triumphant pawn in "Use Me." One or more of the Jonas Brothers was a "sucker for you." Rick James and Teena Marie even collaborated on "I'm a Sucker for Your Love." In love songs, people sing about their struggles with the fool. Luther Vandross sang "Don't Want to Be a Fool."

Love, especially romantic love, shows the fool's game right at the surface. Seduction is itself a kind of gentle scam. Is it okay to post a flattering picture on your dating site? To present your best self early in a relationship? How can you tell if she loves you or just wants to live at Pemberley? The negotiation of intimacy is harrowing in part because of that constant lookout for the con in tension with the rich, urgent rewards of human connection.

I remember feeling, in the early days of meeting my husband, that singular combination of besotted and suspicious. He was a senior at NYU and worked as a busboy in a fancy restaurant. We each lived in Brooklyn with roommates, but my setup was nicer. Every once in a while during that period, while I was otherwise falling in love in a clichéd and predictable manner, I would try to negotiate the precise terms of our affection. A few times I tried to make him articulate whether he was interested in *me* or *the things I had to offer*. To his credit, he found this truly inane. (What I had to offer: two part-time jobs; a refrigerator full of leftovers; my sister's station wagon on long-term loan.) A funny, deadpan philosophy major, he pointed out I was asking an essentially circular metaphysical question about the difference between being loved and being used. How would we distinguish the two, and to what end?

Jane Austen and Karl Marx could have told me: I was my sister's Subaru and it was me.

In love, it's easier to see the stakes for what they are. There is an inexorable uncertainty to the terms of our most personal engagements. This same dynamic applies in all kinds of transactions; it's just buried under more layers of social protection and ritual. It's less acceptable to speak plainly about the inescapable vulnerability. But it helps a bit to be frank, with ourselves and each other, because the real question is not *whether* we'll be suckers, but *how*. As moral agents, we get to choose whether to engage by default or with intention, cynically or with integrity.

Maybe this is where the wary have something to learn from parents, especially mothers. Motherhood is a nonstop hustle, in both senses of that word: a deep con and an urgent set of demands. The immediacy of caretaking—the physical, the emotional, the ethical—brings its own propulsive force. It's a sucker's game—but, in the day-to-day, who cares? The decision matrix is filled in when the baby starts smiling, or the toddler spikes a fever, or your middle schooler is making a new friend.

During a difficult time in my adult life, I called my own parents, again, for what felt like the thousandth time in a week. My mother answered. Even I was tired of hearing myself, and I apologized, embarrassed to be racking up ever more unpayable chits.

"Oh, honey," she said. "You don't have any chits with me."

Acknowledgments

I am very grateful to the many people who helped me write and publish this book. First, I am deeply indebted to Jennifer Gates, my agent at Aevitas, and Karen Rinaldi, my editor and publisher at Harper Wave. Thank you for the opportunity of a lifetime.

I benefited enormously from kind, patient, painstaking editorial guidance from Amanda Moon and Thomas LeBien at Moon & Company, to whom I am very grateful. Additionally, thanks to Iliana Cooper for excellent research assistance.

I am thankful to gracious colleagues and friends who were careful and generous readers of early chapters and drafts: Dave Hoffman, Jean Galbraith, Felicia Lin, Jaclyn Marsh, Shaun Ossei-Owusu, and Dorothy Roberts.

Finally, I cannot adequately articulate the gratitude I owe to my family, but I can at least say thank you: Jane Ryan, Ren Wilkinson, Ivy Wilkinson-Ryan, Caleb Furnas, Jasper Furnas, and June Furnas. I have been very, very lucky.

Notes

Introduction

5 breach in moral terms: Omri Ben-Shahar, *Fault in American Contract Law*, (Cambridge: Cambridge University Press, 2010). See especially Steven Shavell, "Why Breach of Contract May Not Be Immoral Given the Incompleteness of Contracts," 257–70, and Richard Posner, "Let Us Never Blame a Contract Breaker," 3–19.

5 higher than the law would normally allow: Tess Wilkinson-Ryan and Jonathan Baron, "Moral Judgment and Moral Heuristics in Breach of Contract," *Journal of Empirical Legal Studies* 6, no. 2 (2009): 405–23, https://doi.org/10.1111/j.1740–1461.2009.01148.x.

6 "dramaturgical model of social life": Dmitri N. Shalin, "Interfacing Biography, Theory and History: The Case of Erving Goffman," *Symbolic Interaction* 37, no. 1 (2013): 2–40, https://doi.org/10.1002/symb.82.

6 "On Cooling the Mark Out: Some Aspects of Adaptation to Failure": Erving Goffman, "On Cooling the Mark Out," *Psychiatry* 15, no. 4 (1952): 451–63, https://doi.org/10.1080/00332747.1952.11022896.

7 "taken for granted": Ibid., 451.

7 "art of consolation": Ibid., 452.

7 "impossible situation": Ibid., 456.

9 "security and status": Ibid., 451.

9 civil rights activist Oscar Brown Jr.: Leinz Vales, "Trump Twisting Meaning of 'the Snake' Lyrics, Say Oscar Brown Jr.'s Daughters," CNN, February 27, 2018, https://www.cnn.com/2018/02/27/politics/the-snake-africa-oscar-brown-jr-daughters-trump-don-lemon-cnntv/index.html.

9 written as a repudiation of racism: "Insects, Floods and 'the Snake': What Trump's Use of Metaphors Reveals," Public Broadcasting Service, October 22, 2019, https://www.pbs.org/wgbh/frontline/article/insects-floods-and-the-snake-what-trumps-use-of-metaphors-reveals/.

Chapter 1: The Fear

20 coined the term "sugrophobia": Kathleen D. Vohs, Roy F. Baumeister, and Jason Chin, "Feeling Duped: Emotional, Motivational, and Cognitive Aspects of Being Exploited by Others," *Review of General Psychology* 11, no. 2 (2007): 127–41, https://doi.org/10.1037/1089–2680.11.2.127.

20 early experiment in social psychology: Samuel Gaertner and Leonard
Bickman, "Effects of Race on the Elicitation of Helping Behavior: The Wrong
Number Technique," *Journal of Personality and Social Psychology* 20, no. 2 (1971):
218–22, https://doi.org/10.1037/h0031681.

21 "They know me over there": Ibid., 219–20.

21 "Free Money! $1 Bills Available Here!": Dan Ariely cited in Vohs,
Baumeister, and Chin. (This article cites D. Ariely, personal communication,
April 20, 2006.)

22 psychological "power of 'free'": Dan Ariely, *Predictably Irrational: The
Hidden Forces That Shape Our Decisions* (New York: Harper Perennial, 2010).

23 *90 percent of people* avoided their table: Dan Ariely cited in Vohs, Baumeister,
and Chin.

23 fewer than one in four people stopped to inquire: Ibid., 133.

28 *knowledge* of what might have been: See, e.g., Marcel Zeelenberg and Rik
Pieters, "Consequences of Regret Aversion in Real Life: The Case of the Dutch
Postcode Lottery," *Organizational Behavior and Human Decision Processes* 93, no. 2
(2004): 155–68, https://doi.org/10.1016/j.obhdp.2003.10.001.

30 $100 to invest: Daniel A. Effron and Dale T. Miller, "Reducing Exposure
to Trust-Related Risks to Avoid Self-Blame," *Personality and Social Psychology
Bulletin* 37, no. 2 (2011): 181–92, https://doi.org/10.1177/0146167210393532.

30 "Tragedy of the Commons": William Forster Lloyd, *Two Lectures on the
Checks to Population: Delivered Before the University of Oxford, in Michaelmas Term
1832* (United Kingdom: S. Collingewood, 1833). The theory was reinvigorated
and elaborated in a 1968 article of the same name: Garrett Hardin, "The Tragedy
of the Commons," *Science* 162, no. 3859 (1968): 1243–48, https://doi.org
/10.1126/science.162.3859.1243.

31 In the Public Goods Game: Robyn M. Dawes, Jeanne McTavish, and
Harriet Shaklee, "Behavior, Communication, and Assumptions About Other
People's Behavior in a Commons Dilemma Situation," *Journal of Personality and
Social Psychology* 35, no. 1 (1977): 1–11, https://doi.org/10.1037/0022-3514
.35.1.1.

33 most contributions are substantial: Ibid., 6–7.

33 donations will get lower and lower with each new round: See, e.g., Tibor
Neugebauer, Javier Perote, Ulrich Schmidt, and Malte Loos, "Selfish-Biased
Conditional Cooperation: On the Decline of Contributions in Repeated Public
Goods Experiments," *Journal of Economic Psychology* 30, no. 1 (2009) 52–60,
https://doi.org/10.1016/j.joep.2008.04.005.

33 the selfish and the sugrophobic: Dawes, McTavish, and Shaklee, supra note
21, 4–5.

34 nor her friends: Ibid., 7.

34 "how much you alienate me!": Ibid., 7.

35 "status anxiety": Alain de Botton, *Status Anxiety* (London: Penguin, 2014).

35 "our needs are ministered to": Ibid., 6.

35 "the place we occupy in the world": Ibid., 10.

35   more "visual attention": Pia Dietze and Eric D. Knowles, "Social Class and
     the Motivational Relevance of Other Human Beings," *Psychological Science* 27,
     no. 11 (2016): 1517–27, https://doi.org/10.1177/0956797616667721.

36   the dark heart of the terror: *Carrie*, United States: United Artists Corp.,
     1976.

37   experimental subjects with EEG caps in 2013: Marte Otten and Kai J. Jonas,
     "Humiliation as an Intense Emotional Experience: Evidence from the Electro-
     Encephalogram," *Social Neuroscience* 9, no. 1 (2013): 23–35, https://doi.org/10.1
     080/17470919.2013.855660.

37   "often stemming at least in part from a sense of inferiority": Maartje Elshout,
     Rob M. Nelissen, and Ilja van Beest, "Conceptualising Humiliation," *Cognition
     and Emotion* 31, no. 8 (2016): 1581–94, https://doi.org/10.1080/02699931.2016.1
     249462.

39   refused sweetness of any kind for weeks after: J. Garcia, D. J. Kimeldorf,
     and E. L. Hunt, "The Use of Ionizing Radiation as a Motivating Stimulus,"
     Psychological Review 68, no. 6 (1961): 383–95, https://doi.org/10.1037/h0038361.

39   the "Sauce-Bearnaise effect": Martin E. Seligman and Joanne L. Hager,
     "Biological Boundaries of Learning: The Sauce-Bearnaise Syndrome," *Psychology
     Today* 6, no. 3 (1972): 59, https://doi.org/10.1037/e400472009–006.

Chapter 2: Weaponization

41   a man from Aroostook County: Robert H. Sitkoff and Jesse Dukeminier,
     Wills, Trusts, and Estates (New York: Wolters Kluwer, 2022).

43   "boss-man's ladder": Dolly Parton, "9 to 5," Nashville: RCA Nashville, 1980.

44   "the country they built": Ta-Nehisi Coates, "The First White President,"
     Atlantic, May 22, 2018, https://www.theatlantic.com/magazine
     /archive/2017/10/the-first-white-president-ta-nehisi-coates/537909/.

44   ghostwritten by a white man, Bill Ayers: Ibid.

45   "You're the puppet": Eyder Peralta, "Watch: On Russia, Trump Tells
     Clinton 'You're the Puppet,'" National Public Radio, October 20, 2016, https://
     www.npr.org/2016/10/19/498626880/watch-on-russia-trump-tells-clinton
     -youre-the-puppet.

45   "morality is for losers": Anne Applebaum, "In Trump's World, Morality Is
     for Losers," *Washington Post*, October 28, 2021, https://www.washingtonpost
     .com/news/global-opinions/wp/2018/06/20/in-trumps-world-morality-is-for
     -losers/.

45   conquest over their wives: Coates, "The First White President."

47   mental model as a schema: Susan T. Fiske and Patricia W. Linville, "What
     Does the Schema Concept Buy Us?," *Personality and Social Psychology Bulletin* 6,
     no. 4 (1980): 543–57, https://doi.org/10.1177/014616728064006.

48   fill in the blank: J. A. Debner and L. L. Jacoby, "Unconscious Perception:
     Attention, Awareness, and Control," *Journal of Experimental Psychology: Learning,
     Memory, and Cognition* 20, no. 2 (1994), 304–317, https://doi.org/10.1037
     /0278–7393.20.2.304.

48   scripts: Mark W. Baldwin, "Relational Schemas and the Processing of Social
     Information," *Psychological Bulletin* 112, no. 3 (1992): 461–84, https://doi.org/10
     .1037/0033–2909.112.3.461.

50   Brooklyn Bridge: Carl Sifakis, *Hoaxes and Scams* (New York: Facts on File,
     1994).

52   Nash equilibrium: John F. Nash, "Equilibrium Points in N-Person Games,"
     Proceedings of the National Academy of Sciences 36, no. 1 (1950): 48–49, https://
     doi.org/10.1073/pnas.36.1.48.

52   secretarial pool: Douglas Rushkoff, *Life Inc.: How the World Became a
     Corporation and How to Take It Back* (London: Vintage Digital, 2011), 151.

52   two prisoners: Morton Deutsch, "Trust and Suspicion,"
     Journal of Conflict Resolution 2, no. 4 (1958): 265–79, https://doi.
     org/10.1177/002200275800200401.

53   defied the prediction: Rushkoff, *Life Inc.*, 151.

53   "the power of the [Prisoner's Dilemma] metaphor": John M. Orbell and
     Robyn M. Dawes, "Social Welfare, Cooperators' Advantage, and the Option of
     Not Playing the Game," *American Sociological Review* 58, no. 6 (1993): 787–800,
     788. https://doi.org/10.2307/2095951.

54   Wall Street Game: Varda Liberman, Steven M. Samuels, and Lee Ross, "The
     Name of the Game: Predictive Power of Reputations versus Situational Labels in
     Determining Prisoner's Dilemma Game Moves," *Personality and Social Psychology
     Bulletin* 30, no. 9 (2004): 1175–85, https://doi.org/10.1177/0146167204264004.

54   "norm of self-interest": Rebecca K. Ratner and Dale T. Miller, "The Norm
     of Self-Interest and Its Effects on Social Action," *Journal of Personality and Social
     Psychology* 81, no. 1 (2001): 5–16, https://doi.org/10.1037/0022–3514.81.1.5.

55   supposedly selfish motives: Dale T. Miller, "The Norm of Self-Interest,"
     American Psychologist 54, no. 12 (1999): 1053–60, https://doi.org/10.1037/0003–
     066x.54.12.1053.

55   "It got me out of the house": Ibid., 1057.

56   "it is better to be truthful and good than to not": *Dirty Rotten Scoundrels*,
     United States: Orion Pictures, 1988.

59   more likely than millionaires to be audited: Jesse Eisinger and Paul Kiel,
     "Why the Rich Don't Get Audited," *New York Times*, May 3, 2019, https://www
     .nytimes.com/2019/05/03/sunday-review/tax-rich-irs.html.

60   "prove that I work": Tressie McMillan Cottom, "I Am Already Behind,"
     essaying (blog), June 3, 2021, https://tressie.substack.com/p/i-am-already
     -behind.

60   Living While Black: Brandon Griggs, "Here Are All the Mundane Activities
     for Which Police Were Called on African-Americans This Year," CNN, Decem-
     ber 28, 2018, https://www.cnn.com/2018/12/20/us/living-while-black-police
     -calls-trnd/index.html.

61   derogation: Simone Browne, *Dark Matters: On the Surveillance of Blackness*
     (Raleigh, NC: Duke University Press, 2015).

62   an officer's domineering rage: *The Ferguson Report: Department of Justice*

Investigation of the Ferguson Police Department, introduction by Theodore M. Shaw (New York: New Press, 2015).

62 "she wanted to videotape": Ibid., 27.

62 "Nobody videotapes *me*": Ibid., 27.

Chapter 3: Flight

66 To spotlight the sucker slack-off effect: Norbert L. Kerr, "Motivation Losses in Small Groups: A Social Dilemma Analysis," *Journal of Personality and Social Psychology* 45, no. 4 (1983): 819–28, https://doi.org/10.1037/0022-3514.45 .4.819.

70 "betrayal aversion": Jonathan J. Koehler and Andrew D. Gershoff, "Betrayal Aversion: When Agents of Protection Become Agents of Harm," *Organizational Behavior and Human Decision Processes* 90, no. 2 (2003): 244–61, https://doi.org/10.1016/S0749-5978(02)00518-6.

70 Originally dubbed the "Investor Game": Joyce Berg, John Dickhaut, and Kevin McCabe, "Trust, Reciprocity, and Social History," *Games and Economic Behavior* 10, no. 1 (1995): 122–42, https://doi.org/10.1006/game.1995.1027.

72 risk of betrayal that the Investors were afraid of: Iris Bohnet and Richard Zeckhauser, "Trust, Risk and Betrayal," *Journal of Economic Behavior & Organization* 55, no. 4 (2004): 467–84, https://doi.org/10.1016/j.jebo.2003 .11.004.

75 "motivating engine": Adam Grant, *Give and Take: Why Helping Others Drives Our Success* (New York: Penguin Books, 2013), 97–98.

75 "living up to their end of the social contract": Brené Brown, *Daring Greatly: How the Courage to Be Vulnerable Transforms the Way We Live, Love, Parent, and Lead* (London: Penguin Life, 2015), 176.

77 interviews she conducted in a Louisiana parish: Arlie Russell Hochschild, *Strangers in Their Own Land: Anger and Mourning on the American Right* (New York: New Press, 2016).

77 "spends money we don't have on kids": Ibid., 35.

77 "government workers in cushy jobs": Ibid.

79 largely disfavor "welfare": Colin Campbell and S. Michael Gaddis, "'I Don't Agree with Giving Cash': A Survey Experiment Examining Support for Public Assistance," *Social Science Quarterly* 98, no. 5 (2017): 1352–73, https://doi.org /10.1111/ssqu.12338

80 "costs skyrocketing": Sarah Kliff, "GOP Legislator Says Healthy People Lead 'Good Lives,' Should Get Cheaper Health Insurance," Vox, May 2, 2017, https://www.vox.com/policy-and-politics/2017/5/2/15514006/mo-brooks -preexisting-conditions.

82 American attitudes to welfare: Martin Gilens, *Why Americans Hate Welfare: Race, Media, and the Politics of Anti-Poverty Policy* (Chicago: University of Chicago Press, 2000).

83 "year after year": Ibid., 2.

83 "sit home and collect benefits": Ibid., 3.

83   "semi-pro bleeding heart": Westbrook Pegler, "Fair Enough," *New York World-Telegram*, January 7, 1938.

86   little academic fable: Victoria Husted Medvec, "What Might Have Been, What Almost Was, What Used to Be: Subjective Determinants of Satisfaction," PhD diss., Cornell University, 1995.

86   "with a grin": Ibid., 81.

86   "attempting to exploit them": Ibid., 118.

87   problem of divorce bargaining: Tess Wilkinson-Ryan and Jonathan Baron, "The Effect of Conflicting Moral and Legal Rules on Bargaining Behavior: The Case of No-Fault Divorce," *Journal of Legal Studies* 37, no. 1 (2008): 315–38, https://doi.org/10.1086/588265.

88   "Explaining Bargaining Impasse": Linda Babcock and George Loewenstein, "Explaining Bargaining Impasse: The Role of Self-Serving Biases," *Journal of Economic Perspectives* 11, no. 1 (1997): 109–126, https://doi.org/10.1257 /jep.11.1.109.

88   "gain an unfair strategic advantage": Ibid., 110.

89   "below the point they view as fair": Ibid.

Chapter 4: Fight

92   "willing to use them": Paul Stenquist, "Road Rage, 'Zipper Merging' and a Stress-Free Path Through Traffic," *New York Times*, July 10, 2021, https://www .nytimes.com/2021/07/10/business/road-rage-zipper-merging.html.

93   "make it topple over onto them": Kathleen D. Vohs, Roy F. Baumeister, and Jason Chin, "Feeling Duped: Emotional, Motivational, and Cognitive Aspects of Being Exploited by Others," *Review of General Psychology* 11, no. 2 (2007): 127–41, https://doi.org/10.1037/1089–2680.11.2.127.

93   Ultimatum Game: Werner Güth, Rolf Schmittberger, and Bernd Schwarze, "An Experimental Analysis of Ultimatum Bargaining," *Journal of Economic Behavior & Organization* 3, no. 4 (1982): 367–88, https://doi.org/10.1016/0167 –2681(82)90011–7.

94   "rational agents" will try to maximize their own welfare: See, e.g., Ariel Rubinstein, "Perfect Equilibrium in a Bargaining Model," *Econometrica* 50, no. 1 (1982): 97, https://doi.org/10.2307/1912531.

94   Most people can predict the outcome: For a cross-cultural study of Ultimatum Game results, see Alvin E. Roth, Vesna Prasnikar, Masahiro Okuno-Fujiwara, and Shmuel Zamir, "Bargaining and Market Behavior in Jerusalem, Ljubljana, Pittsburgh, and Tokyo: An Experimental Study," *American Economic Review* (1991): 1068–95, http://www.jstor.org/stable/2006907. The line graphs on page 1089 show the consistent fall-off of acceptances, across study locations, for offers of less than 20–30 percent of the total.

96   permitted Receivers to send messages: E. Xiao and Daniel Houser, "Emotion Expression in Human Punishment Behavior," *Proceedings of the National Academy of Sciences* 102, no. 20 (2005): 7398–401, https://doi.org/10.1073/pnas .0502399102.

96 "Enjoy it, I know I will": Ibid., 7400.

96 "bargain for better treatment": Michael Bang Petersen, Daniel Sznycer, Leda Cosmides, and John Tooby, "Who Deserves Help? Evolutionary Psychology, Social Emotions, and Public Opinion About Welfare," *Political Psychology* 33, no. 3 (2012): 398, https://doi.org/10.1111/j.1467–9221.2012.00883.x.

97 punish selfish group members: Ernst Fehr and Simon Gächter, "Altruistic Punishment in Humans," *Nature* 415, no. 6868 (2002): 137–40, https://doi.org /10.1038/415137a.

98 "'They're afraid of being killed,' they said": Margaret Atwood, "Writing the Male Character," in *Second Words: Selected Critical Prose* (Boston: Beacon, 1984), 413.

99 accusations . . . of female infidelity often precede intimate partner violence: Marjorie Pichon, Sarah Treves-Kagan, Erin Stern, Nambusi Kyegombe, Heidi Stöckl, and Ana M. Buller, "A Mixed-Methods Systematic Review: Infidelity, Romantic Jealousy and Intimate Partner Violence Against Women," *International Journal of Environmental Research and Public Health* 17, no. 16 (2020): 5682, https://doi.org/10.3390/ijerph17165682.

99 "redemption of honor": Margo I. Wilson and Martin Daly, "Male Sexual Proprietariness and Violence Against Wives," *Current Directions in Psychological Science* 5, no. 1 (1996): 2–7, https://doi.org/10.1111/1467–8721.ep10772668.

99 "the supreme gentleman": "Transcript of Video Manifesto by Suspected UC–Santa Barbara Shooter," *Washington Post*, May 24, 2014, https://www .washingtonpost.com/national/transcript-of-video-manifesto-by-suspected-uc -santa-barbara-shooter/2014/05/24/04da4618-e381–11e3–9743-bb9b59cde7b9 _story.html.

100 "responsibility of even homicidal cuckolds": Margo I. Wilson and Martin Daly, "Male Sexual Proprietariness and Violence Against Wives," *Current Directions in Psychological Science* 5, no. 1 (1996): 2–7.

100 inherently "treacherous": Jane Schneider, "Of Vigilance and Virgins: Honor, Shame and Access to Resources in Mediterranean Societies," *Ethnology* 10, no. 1 (1971): 22, https://doi.org/10.2307/3772796.

100 "make any man go crazy": Tim Reiterman, Jessica Garrison, and Christine Hanley, "Trying to Understand Eddie's Life—and Death," *Los Angeles Times*, October 20, 2002, https://www.latimes.com/archives/la-xpm-2002-oct-20-me -eddie20-story.html.

101 justification for violence against them: Talia Mae Bettcher, "Evil Deceivers and Make-Believers: On Transphobic Violence and the Politics of Illusion," *Hypatia* 22, no. 3 (2007): 43–65, https://doi.org/10.1111/j.1527–2001.2007. tb01090.x.

101 "accusations of deception": Ibid., 47.

101 "'Come really close so I can terrorize you'": Ivan Natividad, "Why Is Anti-Trans Violence on the Rise in America?," *Berkeley News*, November 22, 2021, https://news.berkeley.edu/2021/06/25/why-is-anti-trans-violence-on-the-rise -in-america/.

102 "very different rules": Greg Sargent, "That Wrenching Video Alone Makes an Utterly Damning Case Against Trump," *Washington Post*, February 10, 2021, https://www.washingtonpost.com/opinions/2021/02/09/that-wrenching-video-alone-makes-an-utterly-damning-case-against-trump/.

102 "the mob is ever the Shape of Fear": Du Bois, W. E. Burghardt, "The Shape of Fear," *North American Review* 223, no. 831 (1926): 294–95, http://www.jstor.org/stable/25110229.

103 "counter efforts of all civil power": Fon Louise Gordon, *Caste & Class: The Black Experience in Arkansas, 1880–1920* (Athens, GA: University of Georgia Press, 2007), 50.

103 "through lies and slander": Adolf Hitler, *Mein Kampf (My Struggle)* (London: Hurst & Blackett, 1938).

103 "the most reprehensible manipulations": Adolf Hitler, transcript of speech delivered at the Reichstag, Berlin, Germany, January, 30, 1939, https://www.ushmm.org/learn/timeline-of-events/1939–1941/hitler-speech-to-german-parliament.

104 a reader who offered this conundrum: Cited in Tess Wilkinson-Ryan, "Breaching the Mortgage Contract: the Behavioral Economics of Strategic Default," *Vanderbilt Law Review* 64, no. 5 (2011): 1547.

106 negligence by a stranger: Tess Wilkinson-Ryan and Jonathan Baron, "Moral Judgment and Moral Heuristics in Breach of Contract," *Journal of Empirical Legal Studies* 6, no. 2 (2009): 405–423, https://doi.org/10.1111/j.1740–1461.2009.01148.x.

106 emotional substrate of their responses: Tess Wilkinson-Ryan and David A. Hoffman, "Breach Is for Suckers," *Vanderbilt Law Review* 63, no. 4 (2010): 1001.

107 "'takes back' the demeaning message": Jean Hampton, "An Expressive Theory of Retribution," in *Retributivism and Its Critics: Papers of the Special Nordic Conference Held at the University of Toronto, 25–27 June 1990* (Germany: Steiner, 1992), 13.

107 responses to a criminal fraud: Kenworthey Bilz, "Testing the Expressive Theory of Punishment," *Journal of Empirical Legal Studies* 13, no. 2 (2016): 358–92, https://doi.org/10.1111/jels.12118.

108 *Bailey v. State of Alabama*: 219 U.S. 219 (1910).

109 rank disrespect: See, e.g., Brittany Farr, "Breach by Violence: The Forgotten History of Sharecropper Litigation in the Post-Slavery South," *UCLA Law Review* 69 (forthcoming 2022).

110 distribution of audits in America: Jesse Eisinger and Paul Kiel, "Who's More Likely to Be Audited: A Person Making $20,000—or $400,000?," ProPublica, December 12, 2018, https://www.propublica.org/article/earned-income-tax-credit-irs-audit-working-poor.

111 "increase the rate of audits": Dorothy A. Brown, "The IRS Is Targeting the Poorest Americans," *Atlantic*, July 27, 2021, https://www.theatlantic.com/ideas/archive/2021/07/how-race-plays-tax-policing/619570/.

111 "its obligations of distributive justice": Aditi Bagchi, "Distributive Justice

and Contract," in *Philosophical Foundations of Contract Law* (Oxford, UK: Oxford University Press, 2014), 193–212, https://doi.org/10.1093 /acprof:oso/9780198713012.003.0011.

Chapter 5: Racial and Ethnic Stereotyping

113 relational contract theory: Ian R. Macneil, "Contracts: Adjustment of Long-Term Economic Relations Under Classical, Neoclassical, and Relational Contract Law," *Northwestern University Law Review* 72, no. 6 (1978): 854.

114 the firm handshake, not the fine print: Stewart Macaulay, "Non-Contractual Relations in Business: A Preliminary Study," *American Sociological Review* 28, no. 1 (1963): 55–67, https://doi.org/10.2307/2090458.

114 "the same country club": Ibid., 63.

115 "The Pain of Word Bondage": Patricia J. Williams, "The Pain of Word Bondage," in *The Alchemy of Race and Rights* (Cambridge, MA: Harvard University Press, 1991), 146–65.

115 "ideal arm's-length transactor": Ibid., 146–47.

116 "sufficient *rights* to manipulate commerce": Ibid., 147–48.

116 "accused unfairly of theft": Lawrence Otis Graham, "I taught my black kids that their elite upbringing would protect them from discrimination. I was wrong," *Washington Post*, November 6, 2014, https://www.washingtonpost.com/post everything/wp/2014/11/06/i-taught-my-black-kids-that-their-elite-upbringing -would-protect-them-from-discrimination-i-was-wrong/.

117 interview with the *Harvard Crimson*: Kathleen Cronin, "Jim Sidanius," *Crimson*, September 14, 2017, https://www.thecrimson.com/article/2017/9/14 /fifteen-professors-2017-jim-sidanius/.

117 intergroup oppression: Jim Sidanius and Felicia Pratto, *Social Dominance: An Intergroup Theory of Social Hierarchy and Oppression* (Cambridge, UK: Cambridge University Press, 2001).

118 "capable of constructing": Ibid., 33.

118 "social hierarchy": Ibid., 38.

118 "moral and intellectual justification": Ibid., 45.

119 beliefs constitute intergroup prejudice: Susan T. Fiske, Amy J. C. Cuddy, and Peter Glick, "Universal Dimensions of Social Cognition: Warmth and Competence," *Trends in Cognitive Sciences* 11, no. 2 (2007): 77–83, https://doi .org/10.1016/j.tics.2006.11.005.

120 a table like this: Amy J. Cuddy, Susan T. Fiske, and Peter Glick, "The Bias Map: Behaviors from Intergroup Affect and Stereotypes," *Journal of Personality and Social Psychology* 92, no. 4 (2007): 631–48, https://doi.org/10.1037/0022 –3514.92.4.631.

121 "worked the miracle": James Sullivan, *History of New York State: 1523–1927* (New York: Lewis Historical Publishing Company, 1927), cited in Peter Francis, "The Beads That Did 'Not' Buy Manhattan Island," *New York History* 78, no. 4 (1997): 411–28, http://www.jstor.org/stable/43460452.

121 "delight": James Grant Wilson, *Memorial History of the City of New-York*

and the Hudson River Valley: From Its First Settlement to the Year 1892 (New York: New York Historical Co., 1892), cited in Francis, "The Beads That Did Not Buy Manhattan Island."

122 naïve and gullible: Camilla Townsend, "Burying the White Gods: New Perspectives on the Conquest of Mexico," *American Historical Review* 108, no. 3 (2003): 659–87, https://doi.org/10.1086/529592.

122 basis of incompetence: Tiane L. Lee and Susan T. Fiske, "Not an Outgroup, Not Yet an Ingroup: Immigrants in the Stereotype Content Model," *International Journal of Intercultural Relations* 30, no. 6 (2006): 751–68, https://doi.org/10.1016 /j.ijintrel.2006.06.005.

122 "popular experience with thieving Gypsies": Louis E. Jackson and C. R. Hellyer, *A Vocabulary of Criminal Slang: With Some Examples of Common Usages* (Portland, OR: Modern Printing Co., 1914), 41.

122 "welsh on a deal": Macaulay, "Non-Contractual Relations in Business," 63.

122 racetrack bets: "Peer Apologises for Using Term 'Welching' in Lords Debate," BBC News, June 23, 2015, https://www.bbc.com/news/uk-politics -33238925.

122 insist on a bargain: William Safire, "Slur Patrol," *New York Times*, July 4, 1993, https://www.nytimes.com/1993/07/04/magazine/on-language-slur -patrol.html.

122 "deceitful": Samuel Oppenheim, *The Early History of the Jews in New York, 1654–1664: Some New Matter on the Subject* (New York: American Jewish Historical Society, 1909), 5.

123 "dangerously anti-Semitic prejudice": Susan T. Fiske, Amy J. Cuddy, Peter Glick, and Jun Xu, "A Model of (Often Mixed) Stereotype Content: Competence and Warmth Respectively Follow from Perceived Status and Competition," *Journal of Personality and Social Psychology* 82, no. 6 (2002): 878–902, https://doi .org/10.1037/0022–3514.82.6.878.

123 funding of the protestors: Associated Press, "As George Floyd Protests Swept the Country, so Did George Soros Conspiracy Theories," *Los Angeles Times*, June 22, 2020, https://www.latimes.com/world-nation/story /2020-06-22/george-soros-conspiracy-theories-surge-amid-george-floyd -protests.

123 Funding migrant caravans: Joel Achenbach, "A Conspiracy Theory About George Soros and a Migrant Caravan Inspired Horror," *Washington Post*, October 29, 2018, https://www.washingtonpost.com/national/a-conspiracy-theory -about-george-soros-and-a-migrant-caravan-inspired-horror/2018/10/28/52df5 87e-dae6–11e8-b732–3c72cbf131f2_story.html.

123 "even annihilation": Amy Cuddy, "The Psychology of Anti-Semitism," *New York Times*, November 3, 2018, https://www.nytimes.com/2018/11/03/opinion /sunday/psychology-anti-semitism.html.

124 core tenets of anti-Asian bigotry: Monica H. Lin, Virginia S. Kwan, Anna Cheung, and Susan T. Fiske, "Stereotype Content Model Explains Prejudice for an Envied Outgroup: Scale of Anti-Asian American Stereotypes," *Personality and*

Social Psychology Bulletin 31, no. 1 (2005): 34–47, https://doi.org/10.1177/014 6167204271320.

124 "achieve too much": Ibid., 37.

124 "model minority": David Crystal, "Asian Americans and the Myth of the Model Minority," *Social Casework* 70, no. 7 (1989): 405–13, https://doi.org/10 .1177/104438948907000702.

124 spike in anti-Asian harassment: "More than 9,000 Anti-Asian Incidents Have Been Reported Since the Pandemic Began," NPR, August 12, 2021, https://www .npr.org/2021/08/12/1027236499/anti-asian-hate-crimes-assaults-pandemic -incidents-aapi.

125 threat of revolt or escape: See, e.g., John W. Blassingame, *The Slave Community: Plantation Life in the Antebellum South* (New York: Oxford University Press, 1972).

125 "sounds like a choice": Elias Leight, "Kanye West Says 400 Years of Slavery 'Sounds Like a Choice,'" *Rolling Stone*, June 25, 2018, https://www.rollingstone .com/music/music-news/kanye-west-says-400-years-of-slavery-sounds-like-a -choice-628849/.

125 "pacified and pacifying": Michael Twitty, "Aunt Jemima and Uncle Ben Deserve Retirement. They're Racist Myths of Happy Black Servitude," NBCNews.com, June 21, 2020, https://www.nbcnews.com/think/opinion /aunt-jemima-uncle-ben-deserve-retirement-they-re-racist-myths -ncna1231623.

126 "They come up here": Randy Billings, "Lepage in Spotlight for Saying Drug Dealers Impregnate 'White Girls,'" *Press Herald*, November 30, 2017, https:// www.pressherald.com/2016/01/07/lepage-accused-of-making-racist-comment -at-bridgton-meeting/.

126 voter fraud: Alexa Ura, "Texas Court of Criminal Appeals Will Review Crystal Mason's Controversial Illegal-Voting Conviction," *Texas Tribune*, March 31, 2021, https://www.texastribune.org/2021/03/31/crystal-mason-texas-voting -ruling/.

126 school placement fraud: John Nickerson, "Bridgeport Woman Arrested for Registering Son in Norwalk School," *Stamford Advocate*, April 16, 2011, https:// www.stamfordadvocate.com/policereports/article/Bridgeport-woman-arrested -for-registering-son-in-1340009.php.

127 "educational neglect": LaToya Baldwin Clark, "Stealing Education," *UCLA Law Review* 68 (2021): 575.

127 fraud in sports: Laura Wagner, "An Anti-Doping Agent Occupied Serena Williams's Property and Everyone Is Being Squirrelly About It," Deadspin, June 27, 2018, https://deadspin.com/an-anti-doping-agent-occupied-serena -williams-s-propert-1826993294.

127 "gender fraud": Anna North, "'I Am a Woman and I Am Fast': What Caster Semenya's Story Says About Gender and Race in Sports," Vox, May 3, 2019, https://www.vox.com/identities/2019/5/3/18526723/caster-semenya-800 -gender-race-intersex-athletes.

128 another amicable deal: C. Engelbrecht and J. Nielsen, "In the Name of Art, an Artist Pockets $83,000 and Creates Nothing," *New York Times*, October 1, 2021, https://www.nytimes.com/2021/10/01/world/europe/danish-museum -artist-jenns-haaning.html.

129 essay collection: Cathy Park Hong, *Minor Feelings: An Asian American Reckoning* (New York: One World, 2020).

129 "rarely 'gets away with it'": Ibid., 114.

129 "because of who he is": Ibid.

130 higher appraisal figures: See, e.g., Brook Endale, "Home Appraisal Increased by Almost $100,000 After Black Family Hid Their Race," *USA Today*, September 13, 2021, https://www.usatoday.com/story/money/nation-now /2021/09/13/home-appraisal-grew-almost-100-000-after-black-family-hid -their-race/8316884002/.

130 clever study on eBay: Ian Ayres, Mahzarin Banaji, and Christine Jolls, "Race Effects on eBay," *RAND Journal of Economics* 46, no. 4 (2015): 891–917, https:// doi.org/10.1111/1756-2171.12115.

130 worse mortgage terms and higher insurance premiums: Michelle Aronowitz, Edward L. Golding, and Jung Hyun Choi, *The Unequal Costs of Black Homeownership* (Cambridge, MA: Massachusetts Institute of Technology, Golub Center for Finance and Policy, 2020), https://gcfp.mit.edu/wp-content/uploads /2020/10/Mortgage-Cost-for-Black-Homeowners-10.1.pdf.

130 cost of being a Black buyer: Ian Ayres, "Fair Driving: Gender and Race Discrimination in Retail Car Negotiations," *Harvard Law Review* 104, no. 4 (1991): 817, https://doi.org/10.2307/1341506.

132 "can't-pays": Melissa B. Jacoby, "Collecting Debts from the Ill and Injured: The Rhetorical Significance, but Practical Irrelevance, of Culpability and Ability to Pay," *American University Law Review* 51 (2001): 229.

132 it's medical debt: David U. Himmelstein, Deborah Thorne, Elizabeth Warren, and Steffie Woolhandler, "Medical Bankruptcy in the United States, 2007: Results of a National Study," *American Journal of Medicine* 122, no. 8 (2009): 741–46, https://doi.org/10.1016/j.amjmed.2009.04.012.

132 less likely than whites to own a house: Natalie Campisi, "The Black Homeownership Gap Is Worse. Here's What's Being Done," *Forbes*, June 18, 2021, https://www.forbes.com/advisor/mortgages/black-homeownership-gap/.

132 a sample of bankruptcy attorneys: Jean Braucher, Dov Cohen, and Robert M. Lawless, "Race, Attorney Influence, and Bankruptcy Chapter Choice," *Journal of Empirical Legal Studies* 9, no. 3 (2012): 393–429, https://doi.org/10.1111/j.1740 -1461.2012.01264.x.

133 "cheat their way out of debt": 151 Cong. Rec. E. 737, E737, cited in Sara Sternberg Greene, "The Failed Reform: Congressional Crackdown on Repeat Chapter 13 Bankruptcy Filers," *American Bankruptcy Law Journal* 89 (2015): 241–68.

133 "abusive, opportunistic, and strategic actors": A. Mechele Dickerson, "Racial Steering in Bankruptcy," *American Bankruptcy Institute Law Review* 20 (2012): 623–50.

134 Racial resentment: Emmitt Y. Riley and Clarissa Peterson, "I Can't Breathe," *National Review of Black Politics* 1, no. 4 (2020): 496–515, https://doi.org/10.1525/nrbp.2020.1.4.496.

135 "Modern Racism Scale": John B. McConahay, Betty B. Hardee, and Valerie Batts, "Has Racism Declined in America?," *Journal of Conflict Resolution* 25, no. 4 (1981): 563–79, https://doi.org/10.1177/002200278102500401.

135 "have gotten more economically than they deserve": Ibid., 568–70.

135 "a new inequality—at their expense": Michael I. Norton and Samuel R. Sommers, "Whites See Racism as a Zero-Sum Game That They Are Now Losing," *Perspectives on Psychological Science* 6, no. 3 (2011): 215–18, https://doi.org/10.1177/1745691611406922.

Chapter 6: Sexism and Suckerdom

137 Judge Pierce: Randy E. Barnett and Nathan Oman, *Contracts: Cases and Doctrine* (Boston: Aspen Publishing/Wolters Kluwer, 2021).

138 "Davenport's dance emporium": *Vokes v. Arthur Murray, Inc.*, LexisNexis (District Court of Appeal of Florida, Second District 1968).

139 "neither in her life nor in her feet": Ibid.

140 "God knows what": *Charles of the Ritz Distributors Corp. v. Federal Trade Commission* (Circuit Court of Appeals, Second Circuit 1944).

141 "the female role during sex": Laurie A. Rudman and Peter Samuel Glick, *The Social Psychology of Gender How Power and Intimacy Shape Gender Relations* (New York: Guilford Press, 2021), 237.

142 women are easy to mislead: Laura J. Kray, Jessica A. Kennedy, and Alex B. Van Zant, "Not Competent Enough to Know the Difference? Gender Stereotypes About Women's Ease of Being Misled Predict Negotiator Deception," *Organizational Behavior and Human Decision Processes* 125, no. 2 (2014): 61–72, https://doi.org/10.1016/j.obhdp.2014.06.002.

143 "beautiful little fool": Francis Scott Fitzgerald, *The Great Gatsby* (New York: Scribner, 1925).

144 small brains: See, e.g., Susan Sleeth Mosedale, "Science Corrupted: Victorian Biologists Consider 'The Woman Question,'" *Journal of the History of Biology* 11, no. 1 (1978): 1–55, http://www.jstor.org/stable/4330691.

144 Bem Sex-Role Inventory: Sandra L. Bem, "The Measurement of Psychological Androgyny," *Journal of Consulting and Clinical Psychology* 42, no. 2 (1974): 155–62, https://doi.org/10.1037/h0036215.

145 Alzheimer's disease: R. M. Henig, "The Last Day of Her Life," *New York Times*, May 14, 2015, https://www.nytimes.com/2015/05/17/magazine/the-last-day-of-her-life.html.

145 psychological instrument that bears her name: Bem, "The Measurement of Psychological Androgyny."

145 "*both* instrumental and expressive": Ibid., 155.

147 gender norms are a web of mandates and loopholes: Deborah A. Prentice and Erica Carranza, "What Women and Men Should Be, Shouldn't Be, Are Allowed to Be, and Don't Have to Be: The Contents of Prescriptive Gender

Stereotypes," *Psychology of Women Quarterly* 26, no. 4 (2002): 269–81, https://doi.org/10.1111/1471–6402.t01–1–00066.

151 gender prejudice comes in sheep's clothing: Peter Glick and Susan T. Fiske, "The Ambivalent Sexism Inventory: Differentiating Hostile and Benevolent Sexism," *Journal of Personality and Social Psychology* 70, no. 3 (March 1996): 491–512, https://doi.org/10.1037/0022–3514.70.3.491.

152 "more power than men": Ibid., 512.

153 his vigilance about his sexual partner's fidelity: See, e.g., Aaron T. Goetz and Kayla Causey, "Sex Differences in Perceptions of Infidelity: Men Often Assume the Worst," *Evolutionary Psychology* 7, no. 2 (April 2009), 253–63, https://doi.org/10.1177/147470490900700208.

153 logic of misogyny: Kate Manne, *Down Girl: The Logic of Misogyny* (Oxford, UK: Oxford University Press, 2017).

153 "watchword": Kate Manne, "The Logic of Misogyny," Boston Review, July 11, 2016, https://bostonreview.net/forum/kate-manne-logic-misogyny/.

153 "good reputation, fame, or similar": Manne, *Down Girl*, 131.

156 heart-wrenching, infuriating tale: Emily Yoffe, "The Uncomfortable Truth About Campus Rape Policy," *Atlantic*, September 29, 2017, https://www.theatlantic.com/education/archive/2017/09/the-uncomfortable-truth-about-campus-rape-policy/538974/.

157 Texas hometown: Elizabeth Bruenig, "A Survivor's Truth, Hiding in Plain Sight," *Washington Post*, September 19, 2018, https://www.washingtonpost.com/opinions/a-survivors-truth-hiding-in-plain-sight/2018/09/19/c45d3ffa-bc25-11e8-8792-78719177250f_story.html.

160 impostor syndrome: Pauline Rose Clance and Suzanne Ament Imes, "The Imposter Phenomenon in High Achieving Women: Dynamics and Therapeutic Intervention," *Psychotherapy: Theory, Research & Practice* 15, no. 3 (1978): 241–47, https://doi.org/10.1037/h0086006.

Chapter 7: The Cool-Out

163 hundreds of Americans gathered: Meryl Kornfield, "Why Hundreds of QAnon Supporters Showed Up in Dallas, Expecting JFK Jr.'s Return," *Washington Post*, November 4, 2021, https://www.washingtonpost.com/nation/2021/11/02/qanon-jfk-jr-dallas/.

163 not a site of child trafficking: Kate Samuelson, "Pizzagate: What to Know About the Conspiracy Theory," *Time*, December 5, 2016, https://time.com/4590255/pizzagate-fake-news-what-to-know/.

163 mass arrests and executions: S. Sardarizadeh and O. Robinson, "Biden Inauguration Leaves QAnon Believers in Disarray," BBC News, January 21, 2021, https://www.bbc.com/news/blogs-trending-55746304.

163 dates in 2017, 2018, or 2019: "Fact Check—False QAnon Claims About Hilary [*sic*] Clinton Being Taken to Guantanamo Bay," Reuters, March 10, 2021, https://www.reuters.com/article/factcheck-clinton-guantanamo/fact-check-false-qanon-claims-about-hilary-clinton-being-taken-to-guantanamo-bay-idUSL1N2L8lOU.

164 "wasn't the right time": Michael Williams and Catherine Marfin, "QAnon Supporters Gather in Downtown Dallas Expecting JFK Jr. to Reappear," *Dallas News*, November 2, 2021, https://www.dallasnews.com/news/2021/11/02 /qanon-supporters-gather-in-downtown-dallas-expecting-jfk-jr-to-reappear/.

164 "taken for granted": Goffman, "On Cooling the Mark Out."

165 "consolation": Ibid., 452.

167 basement of a Yale laboratory: Stanley Milgram, "Behavioral Study of Obedience," *Journal of Abnormal and Social Psychology* 67, no. 4 (1963): 371, https://doi.org/10.1037/h0040525.

167 Bay of Pigs disaster: Irving L. Janis, *Victims of Groupthink: A Psychological Study of Foreign-Policy Decisions and Fiascoes* (Boston: Houghton Mifflin, 1972).

168 human psychology in light of the Holocaust: Radio 4 in Four, "How the Holocaust Created a New Field of Science: The Science of Evil," BBC, n.d., retrieved May 12, 2022, https://www.bbc.co.uk/programmes/articles/4B9rmwvZwQN45 rckdzQKxp2/how-the-holocaust-created-a-new-field-of-science-the-science-of -evil.

168 "lose meaning if isolated": Solomon E. Asch, *Social Psychology* (New York: Prentice-Hall, 1952), 61.

169 "the moment of decision": Solomon E. Asch, "Opinions and Social Pressure," *Scientific American* 193, no. 5 (1955): 31–35, https://doi.org/10.1038 /scientificamerican1155–31.

171 little lottery game: Sunita Sah, George Loewenstein, and Daylian Cain, "Insinuation Anxiety: Concern That Advice Rejection Will Signal Distrust After Conflict of Interest Disclosures," *Personality and Social Psychology Bulletin* 45, no. 7 (2019): 1099–1112, https://doi.org/10.1177/0146167218805991.

173 they conformed: Asch, "Opinions and Social Pressure."

174 members of a doomsday cult: Leon Festinger, Henry W. Riecken, and Stanley Schachter, *When Prophecy Fails: A Social and Psychological Study of a Modern Group That Predicted the Destruction of the World* (New York: Harper Torchbooks, 1956).

174 pick them up at 4:00 p.m.: Julie Beck, "The Christmas the Aliens Didn't Come," *Atlantic*, January 4, 2016, https://www.theatlantic.com/health/archive /2015/12/the-christmas-the-aliens-didnt-come/421122/.

175 mind-numbingly straightforward task: Leon Festinger and James M. Carlsmith, "Cognitive Consequences of Forced Compliance," *Journal of Abnormal and Social Psychology* 58, no. 2 (1959): 203–10, https://doi.org/10.1037/h00 41593.

177 an artifact of its era: Elliot Aronson and Judson Mills, "The Effect of Severity of Initiation on Liking for a Group," *Journal of Abnormal and Social Psychology* 59, no. 2 (1959): 177–81, https://doi.org/10.1037/h0047195.

178 "study on communication": Alexander H. Jordan and Benoît Monin, "From Sucker to Saint: Moralization in Response to Self-Threat," *Psychological Science* 19, no. 8 (2008): 809–15, https://doi.org/10.1111/j.1467–9280.2008.02161.x.

180 The Uneven Split Study: A. Falk, E. Fehr, and U. Fischbacher, "On the

Nature of Fair Behavior," *Economic Inquiry* 41, no. 1 (2003), 20–26, https://doi
.org/10.1093/ei/41.1.20.

181 The Ambiguous Endowment Study: Werner Guth, Steffen Huck, and Peter
Ockenfels, "Two-Level Ultimatum Bargaining with Incomplete Information: An
Experimental Study," *Economic Journal* 106, no. 436 (1996): 593–604, https://doi
.org/10.2307/2235565.

182 The Auction Study: E. Hoffman, K. McCabe, K. Shachat, and V. Smith,
"Preferences, Property Rights, and Anonymity in Bargaining Games," *Games and
Economic Behavior* 7, no. 3 (1994), 346–80, https://doi.org/10.1006/game
.1994.1056.

183 Chris Consumer: Omri Ben-Shahar and Carl E. Schneider, "More Than You
Wanted to Know," in *More Than You Wanted to Know* (Princeton, NJ: Princeton
University Press, 2014), 95.

184 *auctioned off your possessions?*: D. Walker, "What to Look Out for When
Signing a Contract," NBC12.com, n.d., retrieved May 17, 2022, https://www
.nbc12.com/2019/05/15/what-look-out-when-signing-contract/.

184 examples of harsh consumer deals: Tess Wilkinson-Ryan, "The Perverse
Consequences of Disclosing Standard Terms," *Cornell Law Review* 103 (2017): 117.

185 companies break their deals: Uriel Haran, "A Person-Organization
Discontinuity in Contract Perception: Why Corporations Can Get Away with
Breaking Contracts but Individuals Cannot," *Management Science* 59, no. 12
(2013): 2837–53, https://doi.org/10.1287/mnsc.2013.1745.

185 the just-world hypothesis: Melvin J. Lerner, "The Belief in a Just World,"
in *The Belief in a Just World: A Fundamental Delusion* (New York: Plenum, 1980),
9–30, https://doi.org/10.1007/978-1-4899-0448-5_2.

185 individual stress responses: Melvin J. Lerner and Dale T. Miller, "Just World
Research and the Attribution Process: Looking Back and Ahead," *Psychological
Bulletin* 85, no. 5 (1978): 1030–51, https://doi.org/10.1037/0033–2909.85
.5.1030.

186 *A Fundamental Delusion*: Melvin J. Lerner, "The Belief in a Just World."

187 preserve the "existing social arrangements": John T. Jost, Mahzarin
R. Banaji, and Brian A. Nosek, "A Decade of System Justification Theory:
Accumulated Evidence of Conscious and Unconscious Bolstering of the Status
Quo," *Political Psychology* 25, no. 6 (2004): 881–919, https://doi.org/10.1111
/j.1467–9221.2004.00402.x.

187 "Why Men (and Women) Do and Do Not Rebel": John T. Jost, "Why Men
and Women Do and Don't Rebel," in *A Theory of System Justification* (Cambridge,
MA: Harvard University Press, 2020), 249–74, https://doi.org/10.2307/j.ctv1
3qfw6w.14.

187 "packaged up": Goffman, "On Cooling the Mark Out."

187 critical race theory into schools: Cortney O'Brien, "Liberal Media Is
'Brazenly' Lying Saying CRT Isn't Taught in Virginia Schools: Newsbusters,"
Fox News, November 2, 2021, https://www.foxnews.com/media/newsbusters
-highlights-medias-crt-lies.

187 COVID-19 vaccines: Charles Creitz, "Ingraham: Democrat's COVID Lies Are Unraveling," Fox News, January 11, 2022, https://www.foxnews.com/media/laura-ingraham-democrats-covid-lies.

187 "tell other people to believe it": Brian Stelter, *Hoax: Donald Trump, Fox News, and the Dangerous Distortion of the Truth* (New York: Atria/One Signal, 2020).

Chapter 8: Mothersucker

190 "the most important job in the world": Ann Crittenden, *The Price of Motherhood: Why the Most Important Job in the World Is Still the Least Valued* (New York: Henry Holt, 2002).

190 "skin off a snake": Ibid., 12.

191 "biological clock": Richard Cohen, "The Clock Is Ticking for the Career Woman," *Washington Post*, March 16, 1978, https://www.washingtonpost.com/archive/local/1978/03/16/the-clock-is-ticking-for-the-career-woman/bd566aa8-fd7d-43da-9be9-ad025759d0a4/.

191 "structural and ideological pressures upon women to become mothers": Dorothy E. Roberts, "Motherhood and Crime," *Iowa Law Review* 79 (1993): 95–141.

193 "unquestioning and unenlightened": Adrienne Rich, *Of Woman Born: Motherhood as Experience and Institution* (London: Virago, 1991).

193 "*Vous travaillez pour l'armée, madame?*": Ibid., 24.

194 "consolation prizes": Goffman, "On Cooling the Mark Out."

194 "mother of four": David Simon, Robert F. Colesberry, and Nina Kostroff Noble, *The Wire*, season 5, episode 2, HBO, January 13, 2008.

194 "their imprudent ways": Cynthia Lee Starnes, "Mothers as Suckers: Pity, Partnership, and Divorce Discourse," *Iowa Law Review* 90 (2005), 1513–52.

195 motherhood itself that causes attributions of incompetence: Amy J. Cuddy, Susan T. Fiske, and Peter Glick, "When Professionals Become Mothers, Warmth Doesn't Cut the Ice," *Journal of Social Issues* 60, no. 4 (2004): 701–18, https://doi.org/10.1111/j.0022–4537.2004.00381.x.

195 "three days a week": Ibid., 708.

196 "positive behavioral intentions": Ibid., 711.

199 "you're not eating worms": Bill Watterson, *Calvin and Hobbes* (comic strip), Kansas City, MO: Andrews McMeel Publishing, 1993.

200 worse at math: Jessi L. Smith, Kristin Hawkinson, and Kelli Paull, "Spoiled Milk: An Experimental Examination of Bias Against Mothers Who Breastfeed," *Personality and Social Psychology Bulletin* 37, no. 7 (2011): 867–78, https://doi.org/10.1177/0146167211401629.

200 "I . . . got . . . a robe": Lorne Michaels, producer, *Saturday Night Live*, season 46, episode 9, NBC, New York, NY, December 19, 2020.

201 "finish it off": "Man Who Has It All," Twitter post, July, 27, 2021, 3:00 a.m., https://twitter.com/manwhohasitall/status/1419915459508510720.

202 "'me time'?": "Man Who Has It All," Twitter post, August 4, 2021, 3:00 a.m., https://twitter.com/manwhohasitall/status/1422814562714271744.

202 perceptions of "parenting effectiveness": Judith S. Bridges, Claire Etaugh,

and Janet Barnes-Farrell, "Trait Judgments of Stay-at-Home and Employed Parents: A Function of Social Role and/or Shifting Standards?," *Psychology of Women Quarterly* 26, no. 2 (2002): 140–50, https://doi.org/10.1111/1471 –6402.00052.

203 "punishment of parenting mistakes": Amy S. Walzer and Alexander M. Czopp, "Mother Knows Best So Mother Fails Most: Benevolent Stereotypes and the Punishment of Parenting Mistakes," *Current Research in Social Psychology* 16, no. 12 (January 2011), https://doi.org/2014–37827–001.

204 disproportionate rates: Dorothy E. Roberts, "The Racial Geography of Child Welfare," *Child Welfare* 87, no. 2 (2008): 125–50, https://www.jstor.org/stable /48623038.

204 worked at Little Caesars: Chelsea Simeon, "Mom Found Not Guilty of Charges After Leaving Kids in Liberty Motel While She Worked," WKBN.com, May 26, 2021, https://www.wkbn.com/news/local-news/mom-found-not-guilty -of-child-endangering-charges-after-leaving-kids-alone-in-liberty-motel-while -she-worked/.

204 her shift at McDonald's: Conor Friedersdorf, "Working Mom Arrested for Letting Her 9-Year-Old Play Alone at Park," *Atlantic*, July 16, 2014, https:// www.theatlantic.com/national/archive/2014/07/arrested-for-letting-a-9-year -old-play-at-the-park-alone/374436/.

205 "Consider women on welfare": D. Gauthier, "Political Contractarianism," *Journal of Political Philosophy* 5, no. 2 (1997): 136, https://doi-org.proxy.library .upenn.edu/10.1111/1467–9760.00027.

206 "that mother is Black": Dorothy E. Roberts, "Unshackling Black Mother-hood," *Michigan Law Review* 95, no. 4 (1997): 938, https://doi.org/10.2307 /1290050.

206 "'We're paying for that'": Karen Seccombe, Delores James, and Kimberly Battle Walters, "'They Think You Ain't Much of Nothing': The Social Construction of the Welfare Mother," *Journal of Marriage and the Family* 60, no. 4 (1998): 849, https://doi.org/10.2307/353629.

206 welfare causes motherhood: Charles Murray, "Does Welfare Bring More Babies?," *National Affairs*, Spring 1994, https://www.nationalaffairs.com/public _interest/detail/does-welfare-bring-more-babies.

207 "race and IQ": Eric Turkheimer, Kathryn Paige Harden, and Richard E. Nis-bett, "Charles Murray Is Once Again Peddling Junk Science About Race and IQ," Vox, May 18, 2017, https://www.vox.com/the-big-idea/2017/5/18/15655638 /charles-murray-race-iq-sam-harris-science-free-speech.

207 permitted states to impose family caps: "Text of President Clinton's An-nouncement on Welfare Legislation," *New York Times*, August 1, 1996, https:// www.nytimes.com/1996/08/01/us/text-of-president-clinton-s-announcement -on-welfare-legislation.html.

207 "no Black madonna": Dorothy E. Roberts, "The Value of Black Mothers' Work," *Connecticut Law Review* 26, no. 3 (1993): 871.

207 "corrupt the reproduction process at every stage": Dorothy E. Roberts,

Killing the Black Body: Race, Reproduction, and the Meaning of Liberty (New York: Vintage, 1999).

208 "those children use social services": Leo Ralph Chavez, *The Latino Threat: Constructing Immigrants, Citizens, and the Nation* (Stanford, CA: Stanford University Press, 2013), 175.

208 "singling out [of] Latinas": Leo R. Chavez, "A Glass Half Empty: Latina Reproduction and Public Discourse," *Human Organization* 63, no. 2 (2004): 173–88, https://doi.org/10.17730/humo.63.2.hmk4m0mfey10n51k.

210 suckers (or saints!) of last resort: Claire Cain Miller, "When Schools Closed, Americans Turned to Their Usual Backup Plan: Mothers," *New York Times*, November 17, 2020, https://www.nytimes.com/2020/11/17/upshot/schools -closing-mothers-leaving-jobs.html.

210 "father dis-benefit scheme": U.K. House of Commons Hansard, May 13, 1975; cited by S. J. Lundberg, R. A. Pollak, and T. J. Wales in "Do Husbands and Wives Pool Their Resources? Evidence from the U.K. Child Benefit," *Journal of Human Resources* 32, no. 3 (1997): 463. https://doi.org/10.2307/146179.

211 "(for example, the husband)": Ibid., 464.

Chapter 9: The Sucker and the Self

218 Minimal Effort Game: John B. Van Huyck, Raymond C. Battalio, and Richard O. Beil, "Tacit Coordination Games, Strategic Uncertainty, and Coordination Failure," *American Economic Review* 80, no. 1 (1990): 234–48, https://www.jstor .org/stable/2006745.

218 payout matrix: Ibid., 238.

221 "availability heuristic": Amos Tversky and Daniel Kahneman, "Availability: A Heuristic for Judging Frequency and Probability," *Cognitive Psychology* 5, no. 2 (1973): 207–32, https://doi.org/10.1016/0010–0285(73)90033–9.

221 even doctors: See, e.g., Ping Li, Zi yan Cheng, and Gui lin Liu, "Availability Bias Causes Misdiagnoses by Physicians: Direct Evidence from a Randomized Controlled Trial," *Internal Medicine* 59, no. 24 (2020): 3141–46, https://doi.org /10.2169/internalmedicine.4664–20.

222 "laughed at": P. Waldman, "Trump's Pathological Obsession with Being Laughed At," *The Week*, May 31, 2017, https://theweek.com/articles/702268 /trumps-pathological-obsession-being-laughed.

222 "ripped off": "Trump: China, Other Nations Have Become 'Spoiled' on Trade," Reuters, May 17, 2018, https://www.reuters.com/article/us-usa-trade -china-trump/trump-china-other-nations-have-become-spoiled-on-trade-idUSK CN1II2MD.

223 examination of the famous Margaret Mead quotation: J. Mark Weber and J. Keith Murnighan, "Suckers or Saviors? Consistent Contributors in Social Dilemmas," *Journal of Personality and Social Psychology* 95, no. 6 (2008): 1340–53, https://doi.org/10.1037/a0012454.

228 multi-attribute utility theory (MAUT): George P. Huber, "Methods for Quantifying Subjective Probabilities and Multi Attribute Utilities," *Decision*

Sciences 5, no. 3 (1974): 430–58, https://doi.org/10.1111/j.1540–5915.1974
.tb00630.x.

230 "bootstrapping": Robyn M. Dawes, "The Robust Beauty of Improper Linear
Models in Decision Making," *American Psychologist* 34, no. 7 (1979): 571–82,
https://doi.org/10.1037/0003–066x.34.7.571.

232 "suckers" for getting killed: Jeffrey Goldberg, "Trump: Americans Who
Died in War Are 'Losers' and 'Suckers,'" *Atlantic*, September 3, 2020, https://
www.theatlantic.com/politics/archive/2020/09/trump-americans-who-died-at
-war-are-losers-and-suckers/615997/.

233 he was fired: *Fortune v. National Cash Register Co.*, 373 Mass. 96 (Mass. 1977)
364 N.E.2d 1251 (Supreme Judicial Court of Massachusetts, Norfolk), Casetext,
July 20, 1977, https://casetext.com/case/fortune-v-national-cash-register-co-1.

238 lead to trust: C. R. Rogers and R. C. Sanford, "Client-Centered
Psychotherapy," in H. I. Kaplan and B. J. Sadock, eds., *Comprehensive Textbook of
Psychiatry*, vol. 4 (Baltimore: Williams & Wilkins, 1984), 1382.

Conclusion

241 "check for some things": Roseanna Sommers and Vanessa K. Bohns, "The
Voluntariness of Voluntary Consent: Consent Searches and the Psychology of
Compliance," *Yale Law Journal* 128, no. 7 (2018): 1962.

242 customers from the early 1960s: *Williams v. Walker-Thomas Furniture Co.*,
350 F.2d 445 (United States Court of Appeals for the District of Columbia Circuit,
1965).

Index

About the Author

TESS WILKINSON-RYAN is a professor of law and psychology at the University of Pennsylvania Carey Law School. She has a law degree and a doctorate in psychology, and studies the moral psychology of legal decision-making, teaching courses in contracts, consumer law, and leadership. Wilkinson-Ryan grew up in Maine and now lives in Philadelphia with her husband and their two children.